Probability's Nature and Nature's Probability

(A Call to Scientific Integrity)

By Donald E Johnson

Probability's Nature and Nature's Probability

(A Call to Scientific Integrity)
By Donald E Johnson

Cover photographs © iStockphoto.com/Skip O'Donnell, Henrik Jonsson, David Marchal, and Jarek Szymanski.

Earth image and galaxy spiral image courtesy NASA and reproduced with permission from AURA/STScI. Digital renditions of images produced by AURA/STScI are obtainable royalty-free.

Cover design by Jessie Nilo Design + Illustration.

This book, for scientifically-oriented readers, was originally published in March, 2009, and this update was done in October, 2010. A "Lite" version for non-scientists was released in October, 2009. A "Programming of Life" book that highlights the information and computer aspects of life was released in September, 2010. This book contains much of the updated information from the more recent books.

Originally published March, 2009
Updated October, 2010
ISBN: 1-4392-2862-0
Booksurge Publishing
Charleston, SC and Lexington, KY

Table of Contents

Acknowledgments

The author wishes to thank his wife Kris and his daughter Jessie for their encouragement to write this book and their many suggestions during the creation of the book. Jessie's help with the cover design is also greatly appreciated.

The author thanks the peer professionals who initially reviewed this book and offered invaluable suggestions for improvements and corrections of errors. Those professionals include:

Physics Professor Earl Blodgett, University of Wisconsin - River Falls
Biology Professor Jurgen Ziesmann, Azusa Pacific University
Computer Science Professor Lyle Reibling, Azusa Pacific University
Electrical Engineering Manager Patrick Gustafson, Prairie Technologies
Research Biologist Dr Richard C. Vogt, INPA - Brazil
Mathematics Professor Granville Sewell, University Texas - El Paso
Physician Gregory Viehman, MD

Comments or suggestions for error corrections or improvements sent to don@scienceintegrity.net will be greatly appreciated.

Errata and additions are available at scienceintegrity.net.

Tax deductible donations to make copies of this book available free to students may be made at http://www.ideacenter.org/membership/donation.php specifying "PN&NP books."

Introduction

To help the reader understand why this book was written, it is important to know the author's background. From childhood, the author was extremely interested in science, devouring books on anything scientific. His love of science led to a Ph.D. in Chemistry from Michigan State University. During his education, views of the American Chemical Society's "From Molecules To Man" program were totally accepted. At that time, he believed anyone not accepting the "proven" evolutionary scenario was of the same mentality as someone believing in a flat earth. He willingly confronted anyone doubting the evolutionary scenarios, relying on the "facts" presented during his training to promote those scenarios.

The author worked for ten years as a Senior Research Scientist in the medical and scientific instrument field. The complexity of life came to the forefront during continued research, especially when his research group was involved with recombinant DNA during the late 1970's. After several years as an independent consultant in laboratory automation and other computer fields, he began a 20-year career in university teaching, interrupted briefly to earn a second Ph.D. in Computer and Information Science from the University of Minnesota.

Over time, the author began to doubt the natural explanations that had been so ingrained. It was science, and not his religion, that caused his disbelief in the explanatory powers of undirected nature in a number of key areas including the origin and fine-tuning of mass and energy, the origin of life with its complex functional information content, and the increase in complexity in living organisms. This realization was not reached easily, as he had to admit that he had been duped into believing concepts that were scientifically unfounded. The fantastic leaps of faith required to accept the undirected natural causes in these areas demand a scientific response to the scientific-sounding concepts that in fact have no known scientific basis. Scientific integrity needs to be restored so that ideas that have no methods to test or falsify are not considered part of science. This applies to Biblical Creationism as well as to naturalistic "causes" in the areas mentioned above. For example, one should not be able to get away with stating "it is possible that life arose from non-life by ..." without first demonstrating that it is indeed possible (defined in the nature of probability) using known science. One could, of course, state "it may be speculated that ... ," but such a statement wouldn't have the believability that its author intends to convey by the pseudo-scientific pronouncement.

This book will review the many prevalent scenarios that are widely accepted, but need closer examination of their scientific validity. It will also examine the scientific validity of ID (intelligent design) as a model that can be empirically detected and examined. The usefulness of the ID model for furthering scientific inquiry will also be analyzed. It is important to realize that empirical science does not address issues such as "why" (purpose) or "how" (mechanism), but merely "that." For example, the law of gravity has proven to be extremely accurate for determining intersections of moving objects in 3-dimensional space, such as landing on Mars. Science does not yet know "how" gravity works (gravitons have been proposed, especially as part of string theory), and may never know "why" it works, but science has determined "that" it works (every time). Principles and laws that have been demonstrated to be valid will be employed, discounting speculation that has no proven scientific basis. We will be considering "science as we know it," not "science as we don't know it." We will see that ID fits into the guidelines of the scientific method for investigating phenomena, correcting and integrating previous knowledge, or acquiring new knowledge. ID is based on gathering observable, empirical and measurable evidence, subjecting it to specific principles of reasoning.

This book is written for science-minded people, both professionals and the curious. Some of the details necessary for conveying scientific validity may not be understood and appreciated by every reader. Appendix A explains the math requirements needed. This book is not meant to be a comprehensive coverage of all topics (such a book could never be published as new findings occur daily), but does provide representative references for the topics of interest, especially findings supporting ID. Most of the quotations are from scientists who believe in undirected natural causes as the only "valid" science, so don't take their statements as implying their explicit support of ID, even if their own words may demonstrate ID is reasonable. Quotations, even from ID proponents, are not meant to reflect agreement by this author, but rather to reflect the breadth of ideas being considered to explain the- evidence of design that is empirically detectable. There are many more references than should be necessary, but they are needed to counteract the falsity of statements like *"Darwin's theory is now supported by all the available relevant evidence, and its truth is not doubted by any serious modern biologist"* [Daw82]. (Note: all quotes are italicized so the reader can quickly identify them as quotes, by author and year.) It is important to realize that the lack of scientific basis for undirected naturalism is acknowledged by many scientists.

1 Chance: Possible, Probable, and Feasible

As shown in Appendix A (that explains math needed), any outcome that has a non-zero probability (is not impossible) can happen. By the law of probability, given "enough" trials such an outcome will happen, but "enough" may be out of reach in the real world. Consider the following statements to justify evolution. *"Given so much time, the 'impossible' becomes possible, the possible probable, and the probable virtually certain. One has only to wait: time itself performs the miracles"* [WAL55]. *"Given infinite time, or infinite opportunities, anything is possible"* [Daw96Bp139]. As shown in Appendix A, an impossible outcome cannot happen regardless of the number of trials, and a probable outcome will become virtually certain with unlimited trials. A possible outcome becomes probable when its probability is at least 0.5 since any lower probability makes it more likely not to happen. For example, in rolling a pair of dice, it is probable that the sum will be greater than 6 since the probability is 21/36 (0.583) of that happening. If the probability of an outcome is P, the number of trials (n) for that outcome to become probable is $n = \log_2(0.5)/\log_2(1\text{-}P) = -1/\log_2(1\text{-}P)$.

If a die is rolled 100 times with each roll recorded, a very improbable pattern of digits will result, with $P = 6^{-100}$ $(= 1.5 \times 10^{-78})$ probability for that pattern. People who endorse undirected chance formation of life have made speculation like: this pattern is extremely improbable, but there it is – it happened. In a similar way, life is extremely improbable, but it happened by similar chance processes. The fallacy of this line of reasoning is that in rolling the die, any outcome was acceptable so that each roll had a probability of 1 of being correct. As indicated, increasing the number of trials for a certainty does not reduce its probability, so $P = 1^{100} = 1$. The number of trials to make a repeat of the pattern probable is $-1/\log_2(1\text{-}1.5 \times 10^{-78})$. Using Taylor's expansion of $\log_e(1 + x)$ to solve this yields $n \sim -\log_e(.5)/(1.5 \times 10^{-78}) \sim 4.6 \times 10^{79}$. If a roll can be done in one second, the first improbable pattern occurs after only 100 seconds. The repeat of that pattern becomes probable after 1.4×10^{70}centuries, which is approximately 10^{62} times longer than the oldest estimate of the Universe's age. Note that it is possible to roll the repeat pattern on the very next 100 rolls, but its probability is so small that it would occur while humans live to participate that it may be considered "essentially impossible." Since "impossible" already has a definition of zero probability, and the probability in question isn't 0, a more descriptive term is needed for an outcome that is theoretically possible, but which is so improbable that one needn't account for it in reality.

Dictionaries (e.g -- © Random House, Inc. 2006 and The American Heritage® Dictionary of the English Language, Fourth Edition) give definitions of infeasible as impracticable or unworkable -- *"not capable of being carried out or put into practice."* A feasibility study is used to determine whether a possible action is reasonable in terms of its cost. There are many computer applications that are impractical with the current technology because of the doubling of computational power every two years (Moore's Law [Moo65]). An application that would take eight years on current technology is better left undone for four years since both would finish at about the same time, even without any software improvements. There are computer problems that "explode" with increasing problem size. An example is the traditional solution to the Traveling Salesman Problem (TSP), in which the minimum traversal path is found to visit each city once. The order of run-time complexity of the traditional TSP algorithm (step-by-step solution) is O(n!), which makes using that algorithm infeasible for more than a few cities. If the solution for 10 cities takes 10 seconds, a solution for 15 cities would take over 1000 hours, and a 20-city solution would take 2124 centuries, which is clearly infeasible since the salesman would die long before then. Other computer program infeasibilities have been found in neural networks and various non-linear programming applications [Chi04].

Radioactive decay can serve as an example to illustrate the difference between probability, being probable, and being infeasible. The rate of decay has empirically been found to be proportional to the number of radioactive atoms remaining ($dN/dt = -\lambda N$, where λ is the disintegration constant). Since the half-life, h, of ^{14}C is 5730 years, carbon dating [Arn49] can determine the age of many once-living organisms as ^{14}C spontaneously decays to $^{14}N + \beta^-$ (electron). The probability of a radioisotope atom remaining, P_r, after time t is $(\frac{1}{2})^{t/h}$. Ten grams of 5730 year-old wood contains about 3×10^{12} ^{14}C atoms (n). If disintegrations are observed for 30 seconds, the law of probability determines that each atom has a probability of disintegrating of $1 - (\frac{1}{2})^{30s/h}$ (exponent is 1.66×10^{-10}) or 1.15×10^{-10} from the exponential power series expansion. The expected number of disintegrations is the product of the number of atoms and this probability, or 345 in the 30 seconds. The probability that none of the n ^{14}C atoms will disintegrate in the 30 seconds is P_r^n, or 1.48×10^{-150}, which is clearly infeasible. The probability that all n atoms will decay in 30 seconds is $(1.15 \times 10^{-10})^n$. While this probability can be expressed numerically as $10^{-99390000000000}$, it is difficult to imagine (about the same as consecutive SuperLotto lottery jackpot wins each Wednesday and Saturday for 126 trillion years). Since an outcome becomes probable when $(1 - (\frac{1}{2})^{t/h})n$

exceeds 0.5, the time needed to make at least one disintegration probable is 0.0434 seconds for this example.

Some may reject the term "infeasible" as non-scientific. Perhaps a more acceptable term can be used to describe what is already a scientifically-accepted practice [Wil90]. Outlying outcomes are routinely rejected by scientists because the laws of statistics and probability confirm that those outcomes are infeasible [ERP03]. They may be eliminated from consideration because the probability is too low to confirm that the outcomes represent reality. Some have suggested using Borel's guideline *"to set at 10^{-50} the value of negligible probabilities on the cosmic scale"* [Bor50]. Lloyd [Llo02] estimates that the Universe could contain no more than 10^{90} quanta (Note: many physicists believe that a vacuum is filled with energy "quanta" that exceed the Universe's 10^{80} atoms by billions of times). Those quanta could be involved with no more than 10^{120} operations. Dembski uses 10^{-150} as an infeasibility criterion [Dem99]. Since the fastest chemical reaction known takes 10 femtoseconds (10^{-14} sec) [Zew99], if all 10^{80} atoms of the Universe participated in reactions of that speed for14 billion years, less than 10^{111} reactions would theoretically take place. Since the vast majority of reactions are many orders of magnitude slower, if a scenario requires over 10^{111} reactions to become probable, that scenario is clearly infeasible.

In a recent peer-reviewed paper, Abel notes, *"combinatorial imaginings and hypothetical scenarios can be endlessly argued simply on the grounds that they are theoretically possible. But there is a point beyond which arguing the plausibility of an absurdly low probability becomes operationally counterproductive"* [Abe09U]. He then calculates criteria and notes that the Universal Plausibility Principle (UPP) states that *"definitive operational falsification"* of any chance hypothesis is provided by inequalities based on the probabilistic resources of the Earth, solar system, galaxy, or Universe. If a scenario fails to meet the probability inequality standard, *"the hypothetical notion should be declared to be outside the bounds of scientific respectability. It should be flatly rejected as the equivalent of superstition."* Falsification is a criterion that would show a scientific theory is false if the criterion is shown to be true [Pop63]. See Appendix D for a specific theorem and null hypotheses to falsify.

Abel notes that if $f \times {}^{L}\Omega_{A}/\omega < 1$, a scenario is operationally falsified (infeasible), where f = the number of functional objects/events/scenarios that are known to occur out of all ω possible combinations within the theoretical maximum metric ${}^{L}\Omega_{A}$. The level (L) of the metric may be "q" for quantum or "c" for chemical. The astronomical subset(A) of the metric may be "u" for Universe, "g" for our galaxy, "s" for our solar

system, and "e" for Earth. For example:
${}^{C}\Omega_{e}$ is the metric for the maximum chemical reactions on the Earth = 10^{70}
${}^{C}\Omega_{s}$ is the metric for chemical reactions in the solar system = 10^{85}
Since f/ω is the probability of a particular reaction sequence, if the probability < 10^{-70} with Earth resources, a scenario is falsified scientifically (infeasible). The Universal Plausibility Principle is *"independent of any experimental design and data set. No low-probability plausibility assertion should survive peer-review without subjection to the UPP inequality standard of formal falsification... The application of the Universal Plausibility Principle (UPP) precludes the inclusion in scientific literature of wild metaphysical conjectures that conveniently ignore or illegitimately inflate probabilistic resources to beyond the limits of observational science"* [Abe09U].

It should be noted that, unlike Abel's use, "plausible" is often used to describe scenarios that can't be proven, but fit the dictionary.com definition *"having an appearance of truth or reason; seemingly worthy of approval or acceptance; credible; believable,"* or merriam-webster.com definition *"superficially fair, reasonable, or valuable but often specious."* It may be enlightening to consider dictionary.com plausible clarification when scenarios are described as plausible: *"specious describe that which has the appearance of truth but might be deceptive. The person or thing that is plausible strikes the superficial judgment favorably; it may or may not be true: a plausible argument (one that cannot be verified or believed in entirely). Specious definitely implies deceit or falsehood; the surface appearances are quite different from what is beneath: a specious pretense of honesty; a specious argument (one deliberately deceptive, probably for selfish or evil purposes)."* Unfortunately, many non-provable speculations are described as "plausible," with the intention of making them believable, whereas, in reality they are "specious."

Science needs a reality-check if origins is to be studied as science. In what other area would outcomes be published as science if those outcomes had demonstrable probabilities of less than 10^{-100}? When "it's possible that ..." is used, scientists must verify that the pronouncement is indeed possible using known science. Feasibility also needs to be verified using known scientific principles. The feasibility cut-off may vary depending on whether quantum, physical, or chemical interactions are involved, but there is a point where credulity is stretched beyond the breaking point, making science look "foolish" if persisting in treating such paths as pertinent. Without such safeguards, the public will be misled to believe something is science, as opposed to some scientist's speculation.

2 Mass and Energy: Source and Fine-Tuning

The origin of the mass and energy of the Universe has been a topic of much speculation and has continued to be elusive to known science. There are a number of reasons for this, including the non-repeatability of this historic event and the fact that known natural laws fail to account for it. Appendix B gives a brief overview of some of the speculations on origins, pointing out that all such speculation falls outside science as we know it. Regardless of how mass and energy came into existence, the fine-tuned nature of the Universe can be empirically examined and analyzed using known science. The probability of this fine-tuning resulting from undirected natural processes and from some intelligent agent will be analyzed and compared.

To summarize Appendix B, none of the commonly proposed models for the origin of mass and energy of the Universe is in a state that allows it to be classified as science. None can be proved or disproved by known science since known science cannot account for the origin. Each philosophical or theological belief about the origin involves unprovable assumptions that are not verifiable or falsifiable, and would not be bound by known science. A supernatural belief typically involves powerful being(s) outside the Universe. Quantum fluctuation of "nothing" involves an unseeable eternal "ether" capable of spontaneously generating the Universe. An eternally-existing Universe violates the law of increasing entropy and fails to account for the Universe's increasing rate of expansion. Multi-verse collisions involve innumerable unseeable universes with unseeable dimensions. It is no more scientific to state a "scientific-sounding" natural source than to state a supernatural source. Each model requires "beliefs," as opposed to empirically determined science. Regardless of how the Universe came into existence, its properties can be studied and evaluated using known science.

"Fine-tuning" of the Universe allows life on earth to exist. Incompatibility would result by changing any one of dozens of parameters. It should be noted that the majority of the fine-tuned arguments assume a "big-bang" scenario for the origin of this (current) Universe, regardless of the ultimate source (supernatural, oscillating, quantum tunneling, etc.). The physical constants for weak and strong nuclear forces, electromagnetic and gravitational forces, ratios of forces and electron/proton masses, and properties of neutrons are all extremely critical. The expansion rate, mass, density, and age of the Universe are also critical, as is our position within the solar system, galaxy, and Universe. The Earth's orbit, tilt, rotation, magnetic field, atmosphere, and composition are highly unlikely and yet life-critical. Astrophysicist Paul Davies states

"There is for me powerful evidence that there is something going on behind it all... It seems as though somebody has fine-tuned nature's numbers to make the Universe... The impression of design is overwhelming" [Dav88p203]. Stephen Hawking states concerning the constants of physics: *"The remarkable fact is that the values of these numbers seem to have been very finely adjusted to make possible the development of life... For example, if the electric charge of the electron had been only slightly different, stars would have been unable to burn hydrogen and helium, or else they would not have exploded. It seems clear that there are relatively few ranges of values for the numbers that would allow for development of any form of intelligent life"* [Haw88].

Nobel laureate Steven Weinberg reflects on *"how surprising it is that the laws of nature and the initial conditions of the universe should allow for the existence of beings who could observe it. Life as we know it would be impossible if any one of several physical quantities had slightly different values"* [WeiSA]. *"If we nudge one of these constants just a few percent in one direction, stars burn out within a million years of their formation, and there is no time for evolution. If we nudge it a few percent in the other direction, then no elements heavier than helium form. No carbon, no life. Not even any chemistry. No complexity at all"* [Deu06]. If the strong nuclear force *"were just 2% weaker or 0.3% stronger than it actually is, life would be impossible at any time and any place within the universe"* [Swi90]. *"Small changes in the electric charge of the electron would block any kind of chemistry"* [Bar80]. Theoretical physicist Lee Smolin has calculated the probability of star (including our Sun) formation from random parameters as 10^{-229} [Smo97]. While admitting *"life as we know it on Earth would not exist if several of the parameters of physics were different from their existing values"* [Ste07p146], it has been speculated that there is a *"possibility that an appreciable number of planets exist with conditions that, while unsuitable for our form of life, can support some kind of life"* [Ste07p144]. It is important to realize that no scientific proof is offered that life as we don't know it is "possible" (e.g.– non-zero probability), so such statements amount to wishful thinking and pure speculation, not part of science.

"The small value of the cosmological constant is telling us that a remarkably precise and totally unexpected relation exists among all the parameters of the Standard Model of particle physics, the bare cosmological constant and unknown physics" [Abb91]. This constant needs a precision of one part in 10^{120} [Mic99]. Nima Arkani-Hamed has attempted *"to explain why things that appear to be finely, even heroically, tuned actually are not. One possibility, he said, is that our universe is not*

unique but is only part of a vast 'landscape' of universes. If there are huge numbers of universes, perhaps 10 to the 500th power by one estimate, then it is no great stretch to imagine that at least one of them—ours—wound up having extremely small amounts of observed vacuum energy and a weak force that operates on a scale much smaller than expected." [AAAS05]. Note that "possibility" is used without non-zero probability proof for this non-scientific speculation. In analyzing the precision of the original Big Bang entropy, Penrose calculates *"An accuracy of one part in* $_{10}10^{123}$*...the precision needed to set the universe on its course"* [Pen89].

Hawking and Mlodinow describe how miraculous it is that the laws of physics allow for a Universe that is hospitable for life, in which the Universe has an excess of matter over antimatter and galaxies with stars (with planets) that last billions of years. They attempt to "explain" the observation using "M-theory," which unifies gravity with the other weak and strong nuclear and electromagnetic fundamental forces. They predict seven additional space dimensions [Haw10]. *"Besides the absence of any compelling experimental evidence for M-theory, there is another difficulty — its predictions are far from unique. There are* 10^{500} *different ways to curl up the extra seven dimensions and hide them, and how they curl up determines the fundamental constants and what we four dimensional creatures see as the laws of physics. So even if M-theory is the only theory of everything available, there remain* 10^{500} *possibilities for the laws of physics we observe. Thus, say Hawking and Mlodinow, there is no miracle — inflation plus M-theory equals multiverse. Our special Universe is a selection effect: all possibilities have been tried and we find ourselves in the only kind of inflationary patch that can support our existence... Hawking and Mlodinow argue that negative gravitational potential energies allow something to arise from nothing — but that still begs the question of why there is space, time and M-theory at all"* [Tur10]. It also doesn't explain the origin of gravity, on which their unscientific speculation depends. Furthermore, it doesn't explain the fine-tuning in **OUR** Universe, which is the only one that we can observe.

Perhaps someday science will come up with a verifiable undirected natural model to explain the apparent fine-tuning that is evident, and science should certainly seek such knowledge. At this point it would be incorrect to state that "it's possible that science will come up with an undirected natural solution to this question" since that would assume that non-zero probability of such a solution were proven. Hoyle has stated *"A common sense interpretation of the facts suggests that a superintellect has monkeyed with the physics ... The numbers one calculates from the facts seem to me so overwhelm-*

ing as to put this conclusion almost beyond question" [Hoy81U]. Davies states *"If nature is so 'clever' as to exploit mechanisms that amaze us with their ingenuity, is that not persuasive evidence for the existence of intelligent design behind the universe? If the world's finest minds can unravel only with difficulty the deeper workings of nature, how could it be supposed that those workings are merely a mindless accident, a product of blind chance?"* [Dav84]

When data indicate such extremely unlikely parameters, especially when combining highly improbable parameters, design is a natural conclusion. This doesn't prove design, but in analyzing any other physical artifact that clearly demonstrates even a small fraction of this level of fine-tuning, nobody tries to come up with purely natural scenarios for its cause. Note that tentative acceptance of the possibility of design when the data indicate that is the most reasonable course says nothing about the "how" or "why" of the design. It has even been speculated that our Universe is a simulation designed by some advanced society in another universe [Bos02]. *"Another possibility is an unknown agent intervened in the evolution, and for reasons of its own restarted the universe in the state of low entropy characterizing inflation"* [Dys02]. *"To create a new universe would require a machine only slightly more powerful than the LHC* [Large Hadron Collider]... *The big question is whether that has already happened – is our universe a designer universe? By this, I do not mean a God figure,... there is still scope for an intelligent designer of universes as a whole... If our universe was made by a technologically advanced civilisation in another part of the multiverse, the designer may have been responsible for the Big Bang, but nothing more. ... intelligent designers create enough* [10^{500}?] *universes suitable for evolution, which bud off their own universes... It therefore becomes overwhelmingly likely that any given universe, our own included, would be designed rather than 'natural'"* [Gri10].

Design may spur science to look for more properties that indicate design. Philosophers and others could seek to find meaning in those discoveries. One of the criticisms of ID is that it serves no scientific purpose. Suppose ID had been accepted as a valid scientific model 40 years ago. The recent findings of the fine-turned nature of the Universe would be seen as validation of that model, spurring more interest and funding for extended research. One wonders what scientific insights such research would have revealed.

3 Life

Although there is no universally accepted definition of life [Emm97], it often includes characteristics like metabolism, growth, adaptation, and reproduction. *"The existence of a genome and the genetic code divides living organisms from nonliving matter"* is perhaps the most concise definition of life [Yoc05p3]. This definition includes as (at least once) living organisms those that are sterile (e.g. – mules and worker ants) and acellular organisms (e.g. – viruses, which aren't autonomous). While life uses the laws of chemistry and physics, those laws cannot define or explain life any more than the rules of grammar that were used during the preparation of this book define its content.

Amino acids are the building-blocks of life. Each is an organic molecule that has a carboxylic acid ($-CO_2^-$) and an amine ($-NH_3^+$) group attached to the same α-linkage carbon, which often results in both basic and acidic properties. Life uses only α-linkage and levorotatory ("left-handed" chirality, rotating polarized light counter-clockwise) amino acids. The second carbon has another chemical side-group (or side chain) attached (-H for achiral glycine). There are 20 main amino acids for life, each with a standard 3-letter and a 1-letter abbreviation, usually the first letter(s). The 1-letter abbreviation has great utility when using the NCBI Entrez [Entr] protein database that is cross-linked to the Entrez taxonomy database, and has been used to create "water-marks" for artificial genes [Mad08, Gib10]. The amino acids are: Alanine, Arginine (R), Asparagine (Asn/N), Aspartic acid (D), Cysteine, Glutamic acid (E), Glutamine (Gln/Q), Glycine, Histidine, Isoleucine (Ile), Leucine, Lysine (K), Methionine, Phenylalanine (F), Proline, Serine, Threonine, Tryptophan (Trp/W), Tyrosine (Y), and Valine.

A peptide bond results during a catalyzed dehydration reaction in which the carboxylic group of one acid reacts with the amino group of another. Multiple such reactions link amino acids together into polypeptide chains. Functional polypeptide chains longer than approximately 100 amino acids are referred to as proteins which are synthesized in and used by living cells. Sometimes unusable proteins are constructed. An excerpt from the description of the 2004 Nobel Prize in Chemistry is *"Surprisingly many of the proteins created in the cell are faulty from the start. They must be broken down and rebuilt since they can damage the organism... When the proteins have been hacked to pieces, the cell can use their amino acids to synthesize other proteins. When protein degradation does not function correctly, we can become ill"* [Nob04].

Enzymes are catalytic proteins that have special slots to hold other molecules to make chemical reactions feasible. The slots include two for holding the reacting chemicals, one for the ATP energy source (which also requires an enzyme ATP synthase for its production), and slots for establishing which non-chemically-determined DNA/RNA codon (described shortly) is specified. Most of the over 2,000 enzymes can be recognized by the "-ase" ending. Each enzyme lowers the activation energy of a specific required chemical reaction without ultimately being changed itself. Life both requires and manufactures these enzymes, as well as all the other proteins. Enzyme catalyzed reactions take place in the millisecond time-frame, whereas *"uncatalyzed reactions span a range of at least 19 orders of magnitude"* [Lad03]. The longest known biochemical *"half-time - the time it takes for half the substance to be consumed - is 1 trillion years, 100 times longer than the lifetime of the universe. Enzymes can make this reaction happen in 10 milliseconds...Without catalysts, there would be no life at all... It makes you wonder how natural selection operated in such a way as to produce a protein that got off the ground as a primitive catalyst for such an extraordinarily slow reaction"* [Wol03].

Deoxyribonucleic acid (DNA) contains the genetic information of a cell, including the information for constructing proteins (including enzymes) and RNA (Ribonucleic acid). Information is organized as genes (often one gene per protein). Chemically, DNA is a long polymer of nucleotides, each nucleotide being a deoxyribose sugar molecule, one phosphate group, and one base. Nucleotides are joined by ester bonds so that the sugars and phosphates form the DNA backbone, with the bases sticking out to form hydrogen bonds with a second DNA strand to form a double helix [Wat53]. The four bases used in DNA are adenine (A), cytosine (C), guanine (G) and thymine (T), with base pairing always being G-C or A-T. Uracil (U) usually takes the place of T in RNA. The sequence of the four bases along the backbone encodes the genetic instructions (genes) used in the development and functioning of all known living organisms.

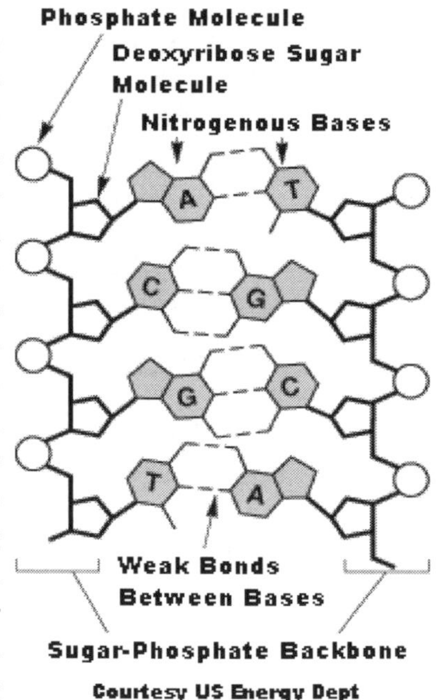

Courtesy US Energy Dept

The genetic code is read by copying stretches of DNA into the related messenger RNA (mRNA) by RNA polymerase during the transcription process. The genetic code consists of three-letter codons formed from a sequence of three nucleotides (e.g. AAG or GAC) that are decoded by a ribosome (complex of RNA and protein) that translates the message sequence by base-pairing the mRNA to transfer RNA (tRNA), and carries the amino acid molecule to the assembly point of the protein.

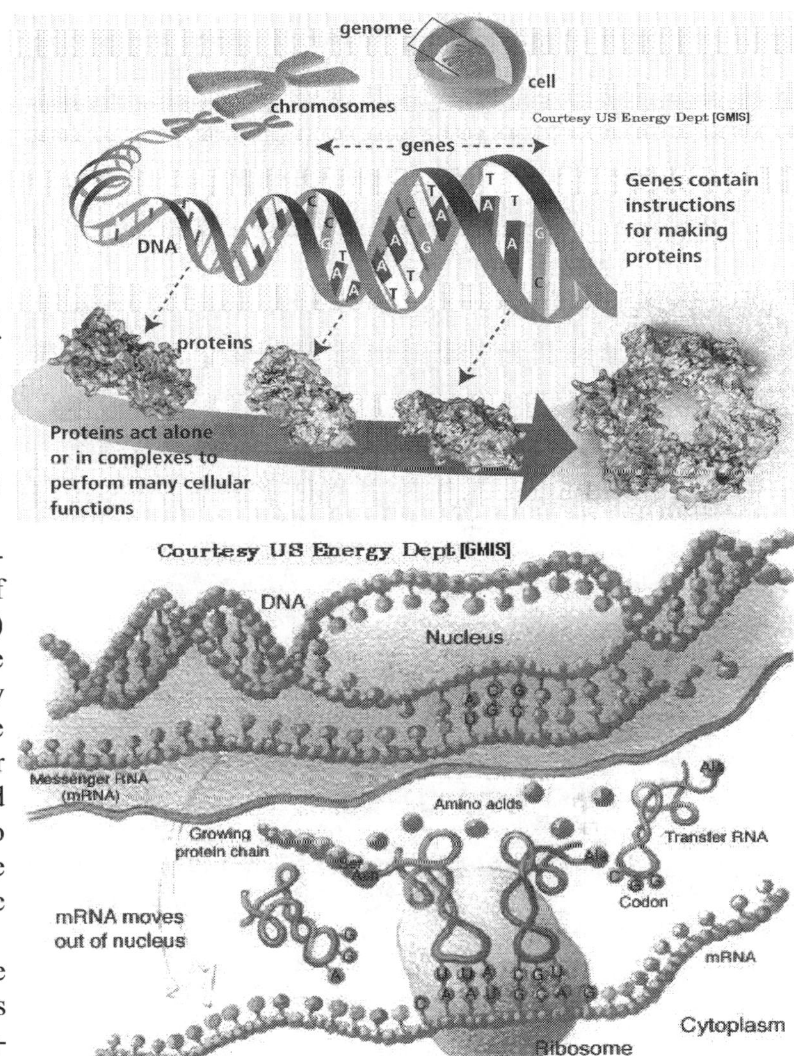

Note that the DNA sequence is digital in base 4 digits, and since each codon is a 3-letter combination, there are 4^3 (= 64) possible codons. These encode the twenty standard amino acids, giving most amino acids more than one possible codon, with codes left over for 3 "stop" codons signifying the end of the gene-coding region space. Non-standard amino acids are sometimes substituted for standard stop codons, depending on the specific mRNA sequence. For example, TGA (UGA in RNA) can code Selenocysteine [Klu97], and TAG (UAG) can code Pyrrolysine [Hao02]. There is considerable speculation on the non-coding parts of DNA intragenic regions known as introns, often dismissed as "junk DNA." The following codons produce the amino acids shown (ATG also starts translation if not in a gene to code Met).

TTT Phe	TCT Ser	TAT Tyr	TGT Cys	ATT Ile	ACT Thr	AAT Asn	AGT Ser
TTC Phe	TCC Ser	TAC Tyr	TGC Cys	ATC Ile	ACC Thr	AAC Asn	AGC Ser
TTA Leu	TCA Ser	TAA Stop	TGA Stop	ATA Ile	ACA Thr	AAA Lys	AGA Arg
TTG Leu	TCG Ser	TAG Stop	TGG Trp	ATG Met*	ACG Thr	AAG Lys	AGG Arg
CTT Leu	CCT Pro	CAT His	CGT Arg	GTT Val	GCT Ala	GAT Asp	GGT Gly
CTC Leu	CCC Pro	CAC His	CGC Arg	GTC Val	GCC Ala	GAC Asp	GGC Gly
CTA Leu	CCA Pro	CAA Gln	CGA Arg	GTA Val	GCA Ala	GAA Glu	GGA Gly
CTG Leu	CCG Pro	CAG Gln	CGG Arg	GTG Val	GCG Ala	GAG Glu	GGG Gly

Helicases (enzyme proteins) are molecular motors that use the chemical energy of ATP to break hydrogen bonds between bases and unwind the DNA double helix into single strands [Tut04]. These enzymes are essential for most processes where enzymes need to access the DNA bases. *"DNA helicases act as critical components in many molecular machineries orchestrating DNA repair in the cell... Multiple diseases including cancer and aging are associated with malfunctions in these enzymes... Helicases are a special category of molecular motors that modify DNA ... by moving along strands of DNA, much the same way cars move on roads, using an energy-packed molecule, adenosine triphosphate (ATP) as a fuel source. Their primary function is to unzip double-stranded DNA, allowing replication and repair of the strands"* [Spi08].

The smallest genome found so far (not autonomous) is in *"the psyllid symbiont Carsonella ruddii, which consists of a circular chromosome of 159,662 base pairs... The genome has a high coding density (97%) with many overlapping genes and reduced gene length"* [Nak06]. Overlapping genes add complexity since *"Normally, transcriptional overlap can interfere with expression of a gene, but these genomes cope with high frequencies of overlap and with termination signals within expressed genes"* [Wil05]. To illustrate the concept, consider the three artificial "proteins" (much shorter than the typical ~300 amino acid length), each having its own pre-coding enabling sequence (not specified here), generated by the following artificial genome sequence.

```
   ...ATGTGTGATGCTACCCTATGTCCAAAAGGGCACCTGCCAATAACCTAGTAGGGTGA...
P1: MetCysAspAlaThrLeuCysProLysGlyHisLeuProIleThr
p2:          MetLeuProTyrValGlnLysGlyThrCysGln
p3:                       MetSerLysArgAlaProAlaAsnAsnLeuValGly
```

Since the DNA code is like computer code in many respects, it is truly amazing that the same "instruction patterns" can perform different overlapping instructions. The author has done this with short sequences (up to eight bytes) of computer code in assembly language, and can assure the reader that it is not trivial to make meaningful operational sequences in which completely different instructions result by starting execution at different locations! *"Overlapping genes are a manifestation of the use of the full information content of the DNA sequence to record and transcribe genetic messages. This phenomenon shows that the source can drive two or even three channels of transcription with appropriate algorithmic instruction"* [Yoc05p92].

The human genome has thousands of overlapping genes. *"However, the origin and evolution of overlapping genes are still unknown"* [Vee04]. Human DNA has a total of approximately 3 billion DNA base pairs and contains an estimated 20,000–25,000 genes [Ste04]. This is down from earlier estimates of over 100,000 genes. *"The Human Genome Sequence Reveals Unexpected Complexity... Only 1.5% of the 3.2 billion base pairs of the human genome encode protein, yet those 31,000 or so genes specify 100,000 to 200,000 distinct proteins"* [Lew06]. *"At least part of the information for the extra proteins may come from the presence of hitherto undiscovered overlapping genes, although more may come from alternative splicing of exons in a single gene"* [Ans-Web]. *"Indeed, as a result of the overlapping genetic messages and different modes of information processing, the specified information stored in DNA is now recognized to be orders of magnitude greater than was initially thought"* [Mey09p462]. The DNA information storage medium is already considerably more complex than any other system known (including anything man-made), and we still don't know the extent of that complexity.

Recently, sub-coded information [Can10] and a second genetic code [Bar10] characterizing alternative splicing have been discovered. Various transcribed RNAs are mixed and matched and spliced into mRNAs for specifying protein construction and other controls, sometimes joining messages that were separated by thousands of nucleotides. MicroRNAs (small RNA segments) regulate large networks of genes by acting as master control switches [Lie10]. Tiny polypeptides (with 11-32 amino acids) can function as "micro-protein" gene expression regulators [Kon10]. Recently, scientists have *"discovered a new language for mRNA, but we have also translated the previously unknown language of up to 17,000 pseudogenes and at least 10,000 long non-coding (lnc) RNAs. Consequently, we now know the function of an estimated 30,000 new entities, offering a novel dimension by which cellular and tumor biology can be regulated, and effectively doubling the size of the functional genome"* [Pan10].

The human genome has 23 pairs of chromosomes, which are organized structures of DNA and proteins found in cells. If stretched out, DNA would be a two-meter chain weighing about 10^{-12} grams. One DNA molecule from each human that has ever lived [Peo02] would have a total mass under 0.1 gram. If all the DNA in a human's body were laid end-to-end, it would be about 50 billion kilometers long [Spe97p30]. DNA is extremely information rich, as Richard Dawkins writes *"Biology is the study of complicated things that give the appearance of having been designed for a purpose... Physics books may be complicated, but ...The objects and phenomena that a physics book describes are simpler than a single cell in the body of its author. And the author consists of trillions of those cells, many of them different from each other, organized with intricate architecture and precision-engineering into a working machine capable of writing a book"* [Daw96Bp1-3].

One might think that the genetic code is "inefficient" in that a codon has 64 possibilities, with only 20 amino acids being specified. It has been discovered that the DNA alphabet is optimum: *"the best of all possible codes"* [Fre00]. When a recently discovered chemical structural parity-bit *"error-coding approach is coupled with chemical constraints, the natural alphabet of A, C, G, and T emerges as the optimal solution for nucleotides"* [Mac06]. This parity error-check can be examined by helicases during unzipping so that structures with errors may be destroyed rather than producing a defective structure. Parity will detect any odd number of errors (with a single error being the most common). The information space in the codon may even provide for Hamming error-correction [Ham50], but more research is needed as the coding for such correction is arbitrary. In most cases, a single nucleotide error will produce the same amino acid when transcribing the gene. Only Methionine (which is also used as the genetic start code) and Tryptophan have no redundant codes. *"The fact that more than one codon is assigned to eighteen of the more common amino acids in protein is seen as very natural, and indeed necessary, to achieve a moderate error-correcting capability in the genetic code"* [Yoc05p42]. Error correction [May04] and repair makes mutations very rare.

A typical cell (except prokaryotic which lacks a membrane-bound nucleus found in an eukaryotic cell) has a nucleus that contains the DNA, whereas the cytoplasm outside the nucleus contains specialized organelles. The lipoprotein cell *"membrane recognizes with its uncanny molecular memory the hundreds of compounds swimming around it and permits or denies passage according to the cell's requirements"* [Bor73].

The organelles include mitochondria that provide energy, the protein-producing factories called ribosomes, golgi that package and store the manufactured proteins, and lysosomes that dispose of waste. *"If genomic DNA is the cell's planning authority, then the ribosome is its factory, churning out the proteins of life. It's a huge complex of protein and RNA with a practical and life-affirming purpose-catalyzing protein synthesis. Bacterial cells typically contain tens of thousands of ribosomes, and eukaryotic cells can contain hundreds of thousands or even a few million of them"* [Bor07]. A RNA regulatory role that relies on RNAs' ability to communicate with one another dramatically increases known functional genetic information. *"The new findings suggest that nature has crafted a clever tale of espionage such that thousands upon thousands of mRNAs and noncoding RNAs, together with a mysterious group of genetic relics known as pseudogenes, take part in undercover reconnaissance of cellular microRNAs, resulting in a new category of genetic elements which, when mutated, can have consequences for cancer and human disease at large"* [Pol10].

In multicellular organisms, cells are typically differentiated to perform specialized functions. Almost all organisms that can be seen without magnification are multicellular, as are all members of the Plantae and Animalia kingdoms. All organisms begin as a single cell, but multicellular organisms have been programmed to produce cells that have considerably different characteristics, even though the DNA is identical. *"Cell fate is governed not only by the genome, but also by chemical changes to DNA and its associated proteins, a research field called epigenetics. These 'epigenetic'* [epi- means above] *tags are one of the ways that genes get switched on or off in different places at different times, enabling different tissues and organs to arise from a single fertilised egg"* [Kim10]. The **epigenome** causes genes to be expressed differently. For example, a stem cell may change into one of many cell types (muscle, brain, blood vessels, etc.) as it continues to divide, activating some genes while suppressing others. In early embryonic development, a series of chemical signals from different signaling pathways cause a single undifferentiated cell to become a highly specialized complex organism with a variety of different cell types arranged in very precise patterns. These patterns, ensure that all the body structures develop correctly, each in the appropriate place. Development uses a precisely regulated interplay of different cell types. *"It is fascinating how the genetic programme of an organism is able to produce such different cell types out of identical precursor cells"* [Sch10]. *"Three neurexin genes can generate over 3,000 genetic messages that help control the wiring of the brain"* [Fre10].

As an example of a complex organ, the human brain contains about 100 billion neuron cells, each linked to as many as 10,000 other neurons (10^{15} total synaptic interconnections). It is estimated that the human brain can perform in excess of 10^{16} operations per second [Hor08], which is more than all the electronic computers in the world put together (although that may not be true much longer due to the advances in computer power and widespread computer use). *"Many neuroscientists do assume that, just as computers operate according to a machine code, the brain's performance must depend on a 'neural code,' a set of rules or algorithms that transforms those* [synaptic electrical] *spikes into perceptions, memories, meanings, sensations, and intentions*" [Hor08]. The brain uses 20% of the body's total oxygen consumption and 15% of the blood flow, even though it is only 2% of the body's weight [Rai02, Cal06]. *"In each of the some 300 trillion cells in every human body, the words of life churn almost flawlessly through our flesh and nervous system at a speed that utterly dwarfs the data rates of all the world's supercomputers. For example, just to assemble some 500 amino-acid units into each of the trillions of complex hemoglobin molecules that transfer oxygen from the lungs to bodily tissues takes a total of some 250 peta operations per second.* [Gil06]. (That's 2.5×10^{17} operations/sec!)

The genome is a unique set of programs embedded in the DNA memory for an organism. *"The great evolutionary biologist George C Williams has pointed out that animals with complicated life cycles need to code for the development of all stages in the life cycle, but they only have one genome with which to do so. A butterfly's genome has to hold the complete information needed for building a caterpillar as well as a butterfly. A sheep liver fluke has six distinct stages in its life cycle, each specialized for a different way of life"* [Daw98p24]. "*The main distinctive features of the living beings are their extreme complexity, which is unmatched in the non-living world, and (not independently) the rather obvious but still overlooked fact that, besides matter and energy, they receive and transmit information, and that life heavily relies on information transfer and conservation. This last point has no equivalent outside the living world and appears as the specific mark which radically differentiates the living world from the non-living one. It makes biology especially relevant to information theory... prompting biologists to use information theory as a main tool*" [Bat07].

Obviously, this chapter just scratches the surface of life. Many books have been written that go into much more detail on all aspects of this extremely complex topic. As more is known about life, the more intricate and complex it becomes. There is no such thing as a "simple" organism. The information in this chapter will serve as a sufficient backdrop for the rest of this book.

4 The Origin of Life

The formation of a living organism from non-living components, abiogenesis, remains a mystery to science. Several scenarios have been proposed, each having characteristics that stretch credulity. This chapter will examine some scenarios that have been (including those now mostly rejected that were once "mainstream") or are currently being examined. There are significant difficulties in trying to model and explain the physical processes (chemistry and physics), but even more problems arise when trying to explain the information content present in all life, which will be covered in the next chapter. Even a "simple" self-reproducing organism contains vast quantities of complex, functional information. Mycoplasma genitalium has a small genome for a free-living organism, containing 482 genes in 580,000 bases. *"Of course, these genes are only functional with pre-existing translational and replicating machinery, a cell membrane, etc. But Mycoplasma can only survive by parasitizing more complex organisms, which provide many of the nutrients it cannot manufacture for itself. So evolutionists must posit a more complex first living organism with even more genes"* [Fra95]. The Venter Institute has found that only 382 of the 482 genes are essential [Gla06]. One estimated lower limit on the minimum genes for existing life is 256 [Mus96]. Molecular biologists estimate that between 318 and 562 kilobase pairs of DNA are required to produce the proteins needed to maintain life in a minimally complex single-celled organism [Ita95]. The theoretical lower limit of a genome is unknown, but it has been speculated that the minimum number of genes for an organism may be 100 or less [Mur07C]. *"Our proposed minimal genome is 113 kbp long and contains 151 genes"* [For06].

"Nobody knows how it happened but, somehow, without violating the laws of physics and chemistry, a molecule arose that just happened to have the property of self-copying – a replicator" [Daw96Cp282-3]. If "*Nobody knows how*" it seems presumptuous to know that it *"just happened."* The question *"How did life begin?"* is one of the *"biggest unanswered questions"* in biology [New04]. Although written 20 years ago, it is still true that "*more than 30 years of experimentation on the origin of life in the fields of chemical and molecular evolution have led to a better perception of the immensity of the problem of the origin of life on Earth rather than to its solution"* [Dos88]. *"What creates life out of the inanimate compounds that make up living things? No one knows. How were the first organisms assembled? Nature hasn't given us the slightest hint. If anything, the mystery has deepened over time"* [Eas07].

"The received view, today, is that life is but an extremely complex form of chemistry... The problem of which molecules came first has been the object of countless debates.... What really matters is that spontaneous genes and spontaneous proteins had the potential to evolve into the first cells. This however, is precisely what molecular biology does not support. The genes and proteins of the first cells had to have biological specificity, and specific molecules cannot be formed spontaneously. They can only be manufactured by molecular machines, and their production requires entities like sequences and codes that simply do not exist in spontaneous processes. That is what really divides matter from life. All components of matter arise by spontaneous processes that do not require sequences and codes, whereas all components of life arise by manufacturing processes that do require these entities. It is sequences and codes that make the difference between life and matter. It is semiosis [symbol translation system] *that does not exist in the inanimate world, and that is why biology is not a complex form of chemistry"* [Bar08S].

Although Walther Loeb generated the amino acid glycine from non-organic material in 1913 [Loe13], the now-famous Miller-Urey [Mil53] implementation of the "prebiotic soup theory" [Opa52] usually is credited as the first production of amino acids from inorganic sources. They designed a spark generator in an oxygen-free highly reduced mixture of methane, ammonia, and hydrogen gases to form basic organic monomers. Products included a racemic (both l- and d- forms, useless to life) mixture of amino acids that were immediately removed from the reaction to prevent rapid decomposition [Cai93]. *"We believe that there must have been a period when the earth's atmosphere was reducing, because the synthesis of compounds of biological interest takes place only under reducing conditions"* [Mil74]. Although doubts [Abe66, Sch72, Yoc81, Tha92, Yoc92, Rod99] are expressed that the mixture of gases used in the Miller-Urey experiment truly reflects the atmospheric content of early Earth, *"It has been agreed by most scientists that Earth's atmosphere was reducing or neutral, and that oxygen was not present. The reducing conditions present were required for the synthesis of the building blocks of life... Besides the bases, ribose and phosphate were also needed to create nucleotides. The sugar ribose has not yet been discovered as a major biproduct in any chemical reaction using molecules present in the prebiotic soup. Also, the ribose that is created comes in two forms, and the right-handed form is the only one used in synthesis of nucleotides. Another problem to the building of nucleotides is that phosphates were not present in large amounts in the prebiotic environment"* [Org94]. *"The question of how life emerged on Earth is a subject that has inspired a great deal of myth, both scientific and otherwise... while most agree that the atmosphere was not reducing..., there is no consensus that it was oxidizing either"* [Ber06P].

Note that the reducing conditions had been assumed because they "were required for the synthesis" of the amino acids, not because of empirical data. Although some believe *"the organic soup in the oceans and ponds on early Earth would have been a more favorable place for the origin of life than previously thought"* [Tia05], *"many researchers now hold that the ancient Earth's atmosphere, compared with the earlier view, had more oxygen and less hydrogen – as the atmosphere does today. Amino acids don't form as readily under that condition as they did in the 1953 experiment, and when they do form, they tend to break apart"* [Gor01]. There is no evidence that a Miller-type scenario did or even could have happened in nature, especially to polymerise to form more complex structures or to form a protocell. In addition, the amino acids formed would not survive in the creation medium, and were not the enantiomer (l-chirality) needed by life (ribose is d-chirality, but wasn't present at all, even though required by life). *"It's nice to talk about replicating DNA molecules arising in a soupy sea, but in modern cells this replication requires the presence of suitable enzymes. Furthermore, DNA by itself accomplishes nothing. Its only reason for existence is the information that it carries and that is used in the production of a protein enzyme. At the moment, the link between DNA and the enzyme is a highly complex one, involving RNA and an enzyme for its synthesis on a DNA template; ribosomes; enzymes to activate the amino acids; and transfer-RNA molecules... How, in the absence of the final enzyme, could selection act upon DNA and all the mechanisms for replicating it? It's as though everything must happen at once: the entire system must come into being as one unit, or it is worthless. There may well be ways out of this dilemma, but I don't see them at the moment"* [Sal71]. *"Chemical evolution is broadly regarded as a highly plausible scenario for imagining how life on earth might have begun... what has emerged over the last three decades ... is an alternative scenario which is characterized by destruction, and not the synthesis of life... The undirected flow of energy through a primordial atmosphere and ocean is at present a woefully inadequate explanation for the incredible complexity associated with even simple living systems"* [Tha92p182&186].

"Many investigators now consider nucleic acids to be much more plausible candidates for the first self-replicating molecules. The work of Watson and Crick and others has shown that proteins are formed according to the instructions coded in DNA. But there is a hitch. DNA cannot do its work, including forming more DNA, without the help of catalytic proteins, or enzymes. In short, proteins cannot form without DNA, but

neither can DNA form without proteins. To those pondering the origin of life, it is a classic chicken-and-egg problem: Which came first, proteins or DNA?" [Hor91] *"Anyone trying to solve this puzzle immediately encounters a paradox. Nowadays nucleic acids are synthesized only with the help of proteins, and proteins are synthesized only if their corresponding nucleotide sequence is present. It is extremely improbable that proteins and nucleic acids, both of which are structurally complex, arose spontaneously in the same place at the same time. Yet it also seems impossible to have one without the other. And so, at first glance, one might have to conclude that life could never, in fact, have originated by chemical means"* [Org94]. The intractable DNA/RNA/protein origin problem has led most scientists to abandon the DNA-based as the first life, even though that is the only life that is known.

Panspermia (exobiogenesis) has been suggested by several authors [Arr08, Cri73, Hoy83, Ber02, Ben10] as a way around the problems of generating life with the genetic code using resources of the Earth. It is suggested that either life forms could have formed by natural processes somewhere in the Universe and then been transported here by accident in carbonaceous chondrites [Coo01], or that advanced extraterrestrial beings deliberately sent life to Earth. Crick and Orgel proposed directed panspermia, suggesting that the seeds of life may have been purposely spread by an advanced extraterrestrial civilization, perhaps facing catastrophic annihilation, or hoping to terraform planets for later colonization [Cri73]. Even Richard Dawkins has stated that such intelligent design may be possible [Exp08]. Hoyle was led to the directed panspermia model after calculating the probability of the chance production of the 2,000 enzymes he assumed were needed for life as 10^{-40000} [Hoy83]. Wickramasinghe explained this in a statement prepared for court stating that *"The most significant single difficulty associated* [with] *the neo- Darwinist view of life is that microorganisms are far too complicated. When bacteria were created, or accomplished, or formed as the case might be, it is true to say that 99.99% of the biochemistry of higher life was already discovered. Some 2000 or so enzymes are known to be crucial over a fairly wide spectrum of life ranging from simple micro-organisms all the way up to Man. The variation of amino acid sequences in these enzymes are, on the whole, rather minor. In each enzyme a number of key positions are occupied by almost invariant amino acids. Let us consider how these enzymes sequences could have been derived from a primordial soup containing equal proportions of the 20 biologically important amino acids. At a conservative estimate say 15 sites per enzyme must be fixed to be filled by particular amino acids for proper biological function. The number of trial assemblies* *needed to find this set is easily calculated to be about 10^40,000 —a truly enormous, super astronomical number. And the probability of discovering this set by random shuffling is 1 in 10^40,000... There is also a serious difficulty to understand how any re-shuffling of amino acids could occur at all in the context of a canonical terrestrial-style primeval soup. To link two amino acids together requires the removal of a water molecule and the supply of some 150 times more energy than heat in the Earth's oceans could supply. In the absence of a joining enzyme used by biology or without an excessively large flux of ultraviolet light at the ocean surface, no new arrangements could be achieved. But even if chemical barriers for the linkages are artificially and miraculously removed, the really vast improbability of 1 in 10^40,000 poses a serious dilemma for the whole of evolutionary science. Life could not be an accident, not just on the Earth alone, but anywhere, anywhere at all in the Universe"* [Wic81].

Some dismiss the calculations of Hoyle and Wickramasinghe as too pessimistic since it is unknown what the minimum number of enzymes is and the minimum number of amino acids in each enzyme. A typical enzyme consists of 250-1000 amino acid residues. For a 200 amino acid enzyme, with all positions equally probable (if not equally probable, the Shannon information of chapter 5 could be used, but that is also unknown), the probability of chance formation of each enzyme is 20^{-200} (= 6×10^{-260}, higher if alternative acids or lower if thermodynamics are considered), or 5×10^{-9368} for only 36 such enzymes. For a *"semi-synthetic minimal cell, a protein expression system with a minimal set of pure and specific enzymes is required... consisting of a specified set of 36 enzymes and ribosomes"* [Mur07P]. This minimum set of 36 enzymes was highly specific and intelligently designed, unlike the chance formation of enzymes needed for the first life. It is also important to realize that the formation reactions for the first enzymes would have no catalysts, which would make the reactions infeasibly slow, even if the needed amino acids were available (which has never been proven). Hoyle's calculations were based on the science known at that time and highlight the difficulty of enzyme production, without which life as we know it couldn't exist.

The "RNA world" [Woe67, Cri68, Bad04] has been proposed to circumvent the difficulty of accounting for the DNA/RNA/protein origination, and currently is the most widely accepted undirected natural scenario. In this scenario, RNA functions as both an enzyme and as a replicator [Kru82, Joy98, Sut10]. The discovery of ribozymes (Nobel prize [Nob89] winning discovery that RNA that can function as an enzyme) has spurred much research and speculation for the RNA world. *"The problem of the origin of life is the problem of the origin of the RNA World, and that everything that followed* *is in the domain of natural selection"* [Org04]. The *"Molecular Biologists Dream"* [Joy99] for the origin of the RNA World *"can be strung together from optimistic extrapolations of the various achievements of prebiotic chemistry and directed RNA evolution.* [One could reasonably wonder what is meant by "directed."]... *First we suppose that nucleoside bases and sugars were formed by prebiotic reactions on the primitive Earth and/or brought to the Earth in meteorites, comets, etc. Next, nucleotides were formed from prebiotic bases, sugars, and inorganic phosphates or polyphosphates, and they accumulated in an adequately pure state in some special little 'pool.' A mineral catalyst at the bottom of the pool—for example, montmorillonite—then catalyzed the formation of long single-stranded polynucleotides, some of which were then converted to complementary double strands by template-directed synthesis. In this way a library of double-stranded RNAs accumulated on the primitive Earth. We suppose that among the double-stranded RNAs there was at least one that on melting yielded a (singlestranded) ribozyme capable of copying itself and its complement. Copying the complement would then have produced a second ribozyme molecule, and then repeated copying of the ribozyme and its complement would... lead to an exponentially growing population. In this scenario this is where natural selection takes over"* [Org04].

Orgel points out several remaining problems including the nonenzymatic synthesis, polymerization, and replication of nucleotides to produce RNA capable of exponential growth in the prebiotic environment, and that *"difficulties remain so severe that alternatives to the de novo appearance of RNA on the primitive Earth deserve serious consideration"* [Org04]. *"But where the first RNA came from is a mystery; it's hard to see how the chemicals on early Earth could have combined to form the complicated nucleotides that make up RNA"* [Dav00]. As De Duve (Nobel Prize-winning biochemist) observes *"The problem is not as simple as might appear at first glance. Attempts at engineering – with considerably more foresight and technical support than the prebiotic world could have enjoyed – an RNA molecule capable of catalyzing RNA replication have failed so far"* [DeD95]. He then speculates how it might have happened in spite of the problems. *"Numerous problems exist with the current thinking of RNA as the first genetic material. No plausible prebiotic processes have yet been demonstrated to produce the nucleosides or nucleotides or for efficient two-way nonenzymatic replication"* [Nel00]. As evolutionary biologist John Maynard Smith observed: *"The origin of the* [genetic] *code is perhaps the most perplexing problem in evolutionary biology. The existing translational machinery is at the same time so complex, so universal, and so essential that it is hard to see how it could have come into existence or how life could have existed without it"* [Smi95]. *"The prebiotic synthesis of nucleotides in a sufficiently pure state to support RNA synthesis cannot be achieved using presently known chemistry"* [Org04].

There are many recent scientific studies on ribozymes and interesting results are sure to follow. It has been speculated that *"an earlier genetic system with a structure completely unrelated to RNA 'invented' RNA... Nucleotides or closely related molecules were synthesized and polymerized by the earlier system for some nongenetic function, and that these molecules somehow developed into molecular Frankensteins"* [Org04]. A 2-base ribozyme system of diaminopurine and uracil has been artificially synthesized [Rea02], and simpler alphabets (coding fewer amino acids) have been speculated [Rod06]. A ligase ribozyme that catalyzes the joining of specific regions of two RNA fragments has been synthesized [Rob07]. A ribozyme has been converted via in vitro evolution to a corresponding deoxyribozyme [Pau06]. But it's important to realize *"ribosome creation requires many RNA-modification enzymes that are still unknown"* [Bar08L]. *"There is no evidence that transcription or RNA replication involve ribozyme catalysis... One must recognize that, despite considerable progress, the problem of the origin of the RNA World is far from being solved."* [Org04]. The prebiotic ribonucleotide synthesis would be racemic, which would produce enantiomeric cross-inhibition [Joy84]. Perhaps one of the biggest problems is *"if the RNA World originated de novo on the primitive Earth, it erects an almost opaque barrier between biochemistry and prebiotic chemistry"* [Org04]. Any speculation about such a world would necessarily deal with "science as we don't know it," despite the insights gained by interesting intelligently-designed current experiments.

Mineral surfaces are proposed by many authors [Fer96, Pop97, Sow01, Haz10] as polypeptide-forming mechanisms in scenarios that bypass the difficulty of forming peptide bonds in aqueous solutions. *"Researchers on the origin of life now conclude that rocks and minerals must have played key roles in virtually every phase of life's emergence—they catalyzed the synthesis of key biomolecules; they selected, protected, and concentrated those molecules; they jump-started metabolism; and they may even have acted as life's first genetic system"* [Haz05]. *"Interactions between amphiphiles and minerals on early Earth may have resulted in the encapsulation of a diverse array of mineral particulates with catalytic properties"* [Han07]. However, *"Theories that involve the organization of complex, small molecule cycles such as the reductive citric acid cycle on mineral surfaces make unreasonable assumptions about the catalytic* *properties of minerals"* [Org00]. *"In the presence of formamide, crystal phosphate minerals may act as phosphate donors to nucleosides ... Thus, activated nucleic monomers can form in a liquid non-aqueous environment in conditions compatible with the thermodynamics of polymerization... In prebiotic scenarios biopolymers can be thought of as condensation products of abiotically formed monomers. Polymers (polysaccharides, peptides, and polynucleotides) will not spontaneously form in an aqueous solution from their monomers because of the standard-state Gibbs free-energy change"* [Cos07]. There is no evidence in undirected nature that minerals are now performing the polypeptide catalyzation or other miracles attributed to them in prebiotic times (or the miraculous removal of all water required by some scenarios). Any information in a crystal structure would have to be in irregularities, as the regular structure contains almost no functional information (a topic of the next chapter). This low information capacity, along with a lack of any feasible mechanism to translate such information into nucleotide sequences eliminate minerals as viable precursors of life.

Other speculation (e. g. osmosis-first [Man10] and metabolism-first [Tre09]) of life's origin attempts to look outside the mainstream views. *"The reaction with purine nucleobases is low-yielding and the reaction with the canonical pyrimidine nucleobases does not work at all... difficulties in nucleobase ribosylation can be overcome with directing, blocking, and activating groups on the nucleobase and ribose... These molecular interventions are synthetically ingenious, but serve to emphasize the enormous difficulties that must be overcome if ribonucleosides are to be efficiently produced by nucleobase ribosylation under prebiotically plausible conditions. This impasse has led most people to abandon the idea that RNA might have assembled abiotically, and has prompted a search for potential pre-RNA informational molecules"* [Sut10]. *"Despite thermodynamic, bioenergetic* [metabolism] *and phylogenetic failings, the 81-year-old concept of primordial soup remains central to mainstream thinking on the origin of life... Here we consider how the earliest cells might have harnessed a geochemically created proton-motive* [electrochemical energy] *force and then learned to make their own, a transition that was necessary for their escape from the vents"* [Lan10]. This is basically an "osmosis-first" (chemical migration through membrane) theory of origins.

"The replicator concept is at the core of genetics-first theories of the origin of life, which suggest that self-replicating oligonucleotides [DNA/RNA] *or their similar ancestors may have been the first "living" systems and may have led to the evolution of an RNA world. But problems with the nonenzymatic synthesis of biopolymers and the*

origin of template [as an information source] *replication have spurred the alternative metabolism-first scenario, where self-reproducing and evolving proto-metabolic networks* [the first chemical cycles] *are assumed to have predated self-replicating genes... In sharp contrast with template-dependent replication dynamics, we demonstrate here that replication of compositional* [composome = "bag" of self-replicating chemicals] *information is so inaccurate that fitter compositional genomes cannot be maintained by selection and, therefore, the system lacks evolvability (i.e., it cannot substantially depart from the asymptotic steady-state solution already built-in in the dynamical equations). We conclude that this fundamental limitation of ensemble replicators cautions against metabolism-first theories of the origin of life"* [Vas10]. Without the information in a genome (or equivalent), evolution can't happen, so that composomes are ruled out as the first "life." *"What is essential, therefore, is a reasonably detailed description, hopefully supported by experimental evidence, of how an evolvable family of cycles might operate. The scheme should not make unreasonable demands on the efficiency and specificity of the various external and internally generated catalysts that are supposed to be involved. Without such a description, acceptance of the possibility of complex nonenzymatic cyclic organizations that are capable of evolution can only be based on faith, a notoriously dangerous route to scientific progress"* [Org08].

The problem of accounting for the information content of life is not addressed in any meaningful way by any abiogenesis model, and will be covered in the next chapter. Although some speculate that creation of life artificially is near [Bar08L], all approaches use existing life components for the attempted synthesis, which is invariably intelligently designed. Some of the problems with undirected origin of life models will be highlighted by the following quotes.

"We once thought that the cell, the basic unit of life, was a simple bag of protoplasm. Then we learned that each cell in any life form is a teeming micro-universe of compartments, structures, and chemical agents—and each human being has billions of cells" [Les98p30-31]. *"The unexpected levels of complexity revealed at the molecular level have further strained the concept of the random assembly of a self-replicating system"* [Swe96]. *"Functionally effective proteins have a vanishingly small chance of arising spontaneously in a prebiotic environment"* [Jim04]. *"When discussing organic evolution the only point of agreement seems to be: 'It happened.' Thereafter, there is little consensus, which at first sight must seem rather odd"* [Mor00]. *"Proteins—and nucleic acids—are not simply polymers, but are co-poly-*

mers, and the kinetics and thermodynamics attending the synthesis of copolymers poses stringent constraints for the biogenesis and growth of specific sequences... there are no reliable methods described in the literature to make copolymers of amino acids or nucleotides under prebiotic conditions" [Lui07]. *"All speculation on the origin of life on Earth by chance can not survive the first criterion of life: proteins are left-handed, sugars in DNA and RNA are right-handed"* [Yoc05p119]. *"Unfortunately, the interpretations of the corpus of publications on the origin of life is false. Those experiments are based on a belief that life is just complicated chemistry and that the origin of life, if it could be found, is emergent from organic chemistry"* [Yoc05p147]. *"The likelihood of life having occurred through a chemical accident is, for all intents and purposes, zero. This does not mean that faith in a miraculous accident will not continue. But it does mean that those who believe it do so because they are philosophically committed to the notion that all that exists is matter and its motion... they do so for reasons of philosophy and not science"* [Gan86].

Given the extreme improbability (it may be zero) of forming life from non-life, 10^{-175} is the thermodynamic probability of forming a simple life-compatible protein of 100 amino acids [Tha92p156-157]. The calculated maximum polypeptide that could be expected in 10^{9} years from a pool of pure, activated biological amino acids would only be 49 amino acid residues long [Yoc77]. Although some have said that the equilibrium calculations of Morowitz don't apply to the prebiotic environment, he has estimated that the probability for the chance formation of the smallest, simplest form of living organism known is one in $10^{340,000,000}$ [Mor79]. There is no way to know what the prebiotic environment may have been, and whether or not that environment was at equilibrium conditions, since most systems approach equilibrium exponentially. Two-time Nobel Prize winner Ilya Prigogine noted something never proved incorrect: *"The statistical probability that organic structures and the most precisely harmonized reactions that typify living organisms would be generated by accident, is zero"* [Pri72].

Since there is no known scientific procedure to generate life in the laboratory, let alone by some unknown prebiotic mechanism, one could assume the probability of life from undirected natural causes is zero. What often is assumed is that since life obviously exists and the only allowable mechanism is undirected and natural, it must have occurred that way, despite the improbability. It is important to remember that as far as science knows, the law of biogenesis, life only arises from life, is valid. Any statement that begins "It's possible that life originated from non-life by ..." is misstated from a probability view since non-zero probability has never been proved.

5 The Information Contained in Life

"The question 'How did life originate?' which interests us all, is inseparably linked to the question 'Where did the information come from?' Since the findings of James D. Watson and Francis H. C. Crick, it was increasingly realized by contemporary researchers that the information residing in the cells is of crucial importance for the existence of life. Anybody who wants to make meaningful statements about the origin of life would be forced to explain how the information originated. All evolutionary views are fundamentally unable to answer this crucial question" [Git97p99]. *"There are no chemical bonds between the bases. Thus, there are no chemical rules to determine the order in which the bases will be attached to the background"* [Dav02]. Carl Sagan wrote: *"The information content of a simple cell has been established as around 10^{12} bits, comparable to about a hundred million pages of the Encyclopaedia Britannica"* [Sag97]. *"Due to the abstract character of function and sign systems, life is not a subsystem of natural laws. This suggests that our reason is limited in respect to solving the problem of the origin of life and that we are left accepting life as an axiom... Life express both function and sign systems, which indicates that it is not a subsystem of the universe, since chance and necessity cannot explain sign systems, meaning, purpose, and goals"* [Voi06].

The coded information *"may be compared to a book or to a video or audiotape, with an extra factor coded into it enabling the genetic information, under certain environmental conditions, to read itself and then to execute the information it reads. It resembles, that is, a hypothetical architect's plan of a house, which plan not only contains the information on how to build the house, but which can, when thrown into the garden, build entirely of its own initiative the house all on its own without the need for contractors or any other outside building agents... Thus, it is fair to say that the technology exhibited by the genetic code is orders of magnitude higher than any technology man has, until now, developed. What is its secret? The secret lies in its ability to store and to execute incredible magnitudes of conceptual information in the ultimate molecular miniaturization of the information storage and retrieval system of the nucleotides and their sequences"* [Wil87]. Evolutionary biologist George Williams observed: *"Evolutionary biologists have failed to realize that they work with two more or less incommensurable domains: that of information and that of matter... These two domains will never be brought together in any kind of the sense usually implied by the term 'reductionism.'... Information doesn't have mass or charge or length in millime-*

ters. Likewise, matter doesn't have bytes... This dearth of shared descriptors makes matter and information two separate domains of existence, which have to be discussed separately, in their own terms" [Wil95].

"One cell division lasts from 20 to 80 minutes, and during this time the entire molecular library, equivalent to one thousand books, is copied correctly" [Git97p90]. *"Crick expounded and enshrined what he called the 'Central Dogma' of molecular biology. The Central Dogma shows that influence can flow from the arrangement of the nucleotides on the DNA molecule to the arrangement of amino acids in proteins, but not from proteins to DNA. Like a sheet of paper or a series of magnetic points on a computer's hard disk or the electrical domains in a random-access memory – or indeed all the undulations of the electromagnetic spectrum that bear information through air or wires in telecommunications – DNA is a neutral carrier of information, independent of its chemistry and physics... As the Central Dogma ordains and information theory dictates, the DNA program is discrete and digital, and its information is transferred through chemical carriers – but it is not specified by chemical forces. Each unit of biological information is passed on according to a digital program – a biological code – that is transcribed and translated into amino acids"* [Gil06].

Microsoft founder Bill Gates writes: *"Human DNA is like a computer program but far, far more advanced than any software we've ever created"* [Gat96]. Can you imagine how believable it would be if someone were to suggest that the Windows 7 operating system just arose by natural processes without intelligence? *"A coding system always entails a nonmaterial intellectual process. A physical matter cannot produce an information code. All experiences show that every piece of creative information represents some mental effort and can be traced to a personal idea-giver who exercised his own free-will, and who is endowed with an intelligent mind"* [Git97p07].

This chapter will examine information in general and how it applies to life. Life's information origin, modification, preservation, and detection as well as capacity and content will be covered. It's important to realize that information content is essentially massless, but the information medium has physical qualities. For example, a computer USB flash drive has the capacity to hold a certain quantity of data, but its weight doesn't measurably change by changing the data content. The smallest numeric base that can hold data is binary since a bit (binary digit) can be either 0 or 1. Anything that has only two possibilities can be represented by a bit, including male/female, married/single, resident/non-resident, etc., with an arbitrary assignment of which choice is 1 and which is 0. A byte (8 bits) can hold an array of eight binary (Boolean) data.

A base 4 digit has possible values 0, 1, 2, and 3, which can be expressed by the two-bit binary numbers 00, 01, 10, or 11, or by A, C, G, or T (using an arbitrary alphabet assignment of genetic symbols).

"Information" has three significant meanings that are important when considering the information of life: "functional" [Szo03, Dur07], "Shannon" [Sha48], and "prescriptive" [Abe09P]. Information always involves contingency that rules out other possibilities. Data and Shannon information (probabilistic complexity) may or may not be useful. Functional information (a subset of Shannon information), on the other hand, is useful or meaningful (about something). Prescriptive information is an algorithmic subset (recipe) of functional information. To illustrate the three information types, consider the data typed into a word-processing program. Most such data is functional in that it has a purpose of communicating information to the ultimate reader of that information. If a monkey typed random data into the program, that complex data would have no purpose, but would have a very high Shannon information content since Shannon information deals only with the probability of the data pattern, irrespective of any meaning. A computer program typed into the wordprocessor is more than just functional, but is prescriptive in that it contains instructions to accomplish objectives based on data to be supplied during the execution of the program being typed. Prescriptive information expresses the decisions to be made and the criteria for the different execution paths. A computer problem is formally solved before physically implementing it (a program doesn't just appear on a disk).

To further clarify the difference between data and functional information, a "blank" disk drive has only meaningless data (zeros and ones), but no functional information. If the disk were erased so that all data were zero, there would technically be one bit of Shannon information (plus a quantity if capacity is unknown) which could have been produced by the algorithm (prescriptive problem solution):
For each bit in the capacity(write 0). If the "0" in the algorithm were replaced by "random 0 or 1," there would be no functional (but the maximum Shannon) information recorded since random (chance) data contains no functional information as it is functionally useless. Exceptions to randomness being "nonfunctional" would be games of chance, such as lotteries, which contain pseudo-information in that a winner is declared arbitrarily to be one picking a match to the randomly produced data.

Any rational number (ratio of two integers) has information limited by the repeating portion of the division result. For example, π (pi) can be approximated by 22/7 = 3.142857..., with the underlined digits repeating forever. If additional digits were written, those digits would be data, but would add no information. The exact representation of π cannot be expressed by any finite sequence of numbers since π is not rational. This does not mean that π has infinite information, since π can be calculated to any degree of precision desired by a computer program of finite length. The information in π is no more than the number of bits in such a program.

The difference between data and functional information can be illustrated with the difference between a protein and a simple chemical polymer. A protein has high functional information content since the sequence of its amino acid components is very specific, and would require an algorithm as long as the number of acids to express that information, according to algorithmic information theory [Cha07]. Polyethylene, on the other hand, could have more data than a protein, but contains very little information as its algorithm (information can't exceed the number of bits to implement) could be: **write("H"); repeat write("-CH_2-") until randomly stopped; write("H")**.

Information may not be obvious, as in the following two 29-bit strings:
11011100101110111000100110 and 10101100001110111011101000011.
The first string is counting to ten: 1 10 11 100 101 110 111 1000 1001 1010, whereas the second string was generated by a binary random number generator. If each bit were reversed, the first would still contain the same information, but in an encrypted form, making the information harder to ascertain. Note that the probability of each pattern is $2^{-29} = 1/536870912 = 1.86 \times 10^{-9}$. Note also that the first string isn't necessarily functional information since it may be feasible to generate it by random processes, and random processes can never produce functional information.

Information transfer always requires that both sender and receiver know the transmission protocol. For example, a single bit of information could be conveyed by the protocol according to the color of a blouse following a proposal: red means I'll marry you, blue means I won't. Other clothing colors would be "noise" to be ignored until the meaningful binary datum is communicated. The "one if by land, two if by sea" is also a binary message using two unary (base 1, which can't store information) objects. If the intended receiver fails to receive the information, the message remains information. For example, if an encrypted message is intercepted, considerable time and effort may be spent trying to extract the information, even if never successful (the interceptor wasn't the intended receiver).

The hand can be an example of an information medium. There are over 1000 signs defined by American Sign Language [Ten98], but simpler finger communication is widely used by most people. The most widely used system is the five (or ten) unary object protocol, holding up the number of digits to represent a number. One could use each digit as a binary number so that 1024 ($= 2^{10}$) values could be represented using ten fingers, but most people would find that protocol awkward. In order for meaningful communication to occur, both sender and receiver must know the protocol and meaning. Note that it is not quantity of message that makes information, it is the agreed protocol. That protocol involves arbitrary rules, not laws. Nothing forces a protocol.

The SETI (Search for Extraterrestrial Intelligence) project [SETI] is the largest single (distributed) computer application in the world, executing about 4×10^{13} floating-point operations per second (operations on non-integers). It uses over three million computers connected over the Internet to analyze data (from radio telescopes that listen to narrow-bandwidth radio signals from space) attempting to extract information that might indicate intelligent life is somewhere other than Earth. If one were to detect the previous 29-bit pattern (page 30) within the analyzed signals, one wouldn't publish the finding as proof that there are aliens out there since that pattern is probable approximately every two hours on each computer. If, however, one were to detect the first 100 prime numbers (number divisible only by itself and 1) in the data, the infeasibility (probability is 10^{-233}) of that pattern arising by chance would indicate that intelligence is out there. If data from each of the maximum estimated 7×10^{21} stars [Ast03] were examined every 10^{-9} second over 14 billion years, the probability of even one such pattern occurring by chance is 10^{-183}, which is clearly infeasible. In the movie and book "Contact" [Sag85], intelligence was detected in 25 prime numbers found.

Unambiguously detecting functional information without an agreed protocol requires detecting meaningful data of such quantity that chance is eliminated as its cause. If E.T. sent a one-time sequence of the first ten prime numbers, that sequence would be information from the sender, but couldn't be differentiated from random data that just happened to have the same value as a meaningful message had there been an agreed protocol for transmitting a short prime number sequence. For a small amount of data without an agreed protocol, information cannot be detected; at best, one could detect a pattern that may be information. Once a pattern is known "for certain," more data aren't informational since they are predictable (a certainty isn't information).

Shannon information theory [Sha48] deals with the reduction of possibilities or uncertainty such that the amount of "information" (actually complexity) in a string of symbols is inversely related to the probability of the occurrence of that string. Shannon uncertainty provides only a mathematical measure of improbability, not whether a symbol string is meaningful or significant (no functionality is required).

Shannon information quantifies the data contained in a message as the minimum message length necessary to communicate that data via information "entropy": $\mathbf{H(S) = -\sum_{i=1}^{n} p_i \log_2(p_i)}$, where the source alphabet, $S = \{a_1, \ldots, a_n\}$, has discrete probability distribution $\{p_1, \ldots, p_n\}$ where p_i is the probability of the symbol a_i. This represents the absolute limit on the best possible lossless compression for any storage or communication of that data. For example, archiving programs like Zip can create a shorter file than the original by translating the original message into a message with the same Shannon information, but with a more concise representation using a compression algorithm like Lempel-Ziv [Ziv78]. "Lossless" means that when uncompressed, an exact copy of the original data is obtained. The Shannon information of a message is the minimum bits per symbol (usually not an integer) multiplied by the number of symbols in the message. For example, English text typically has 0.6-1.5 bits per letter [Sha50], which is considerably below the five (or six for case sensitivity, or eight as it's usually stored) bits per letter to represent that text in a computer.

A long string of repeating characters has 0 information as soon as every character is predictable, e.g. in "junkjunkjunk...," only the first "junk" contributes fully to the information. A string of random letters in the range 'A'-'P' would have 4.0 bits per letter since there are 2^4 (= 16) choices with no predictability. The Shannon information is a measure of the average content that is unknown. In a pair of tossed dice, a sum of 2 or 12 has very high information content, since each die would be known for certain (either 1 or 6). A sum of 7, on the other hand, has low information content, since there is considerable uncertainty of the value of each die.

A coin toss has one bit of Shannon information since the result of each toss can only be predicted to 0.5 probability. A 2-headed coin, on the other hand, would very quickly degenerate to 0 additional information since one could predict the result with a high degree of certainty. Note that random data, although it has zero functional information, has the maximum possible probabilistic complexity since there is no

Low Information (less specific): 6 ways to make 7

High Information: 1 way to make 2 or 12

predictability and therefore the "message" cannot be compressed using a more concise alphabet. The Shannon information of this book is higher than that of typical English text since subscripts, superscripts, and special characters are used. In an early version, the Zip program reduced 5 chapters from 85,683 to 33,278 bytes, so the Shannon information was known to be under 266,224 bits (8 bits per byte). If Zip's compression alphabet is double the theoretical minimum, the estimated Shannon information in those 5 chapters was about 133K bits.

Remember that information can be stored only by contingency, that is, the storage digit can be placed arbitrarily into any of its possible values. DNA is an extremely stable storage system. In the DNA sequence, there are strong ester bonds joining the nucleotides, with no chemical determination as to which base should be adjacent to another base. Since each position is totally free to take any base, DNA is an ideal information storage system. If a particular base had a tendency to attach next to another particular base, the information-carrying capacity would be reduced (to zero if bonding were determined by "law"). The stability of the DNA helix, in which two strands wrap around each other, is determined by the bases. Two bases, A and G, have large double-ring (purine) structures, while C and T have smaller single-ring (pyrimidine) structures (see page 10 figure). The hydrogen bond between two bases in adjacent strands always involves a weak bond between one of each sized molecule, so that bonds are always A-T or C-G between the helix strands. This results in a constant distance between the weakly-bonded bases to produce a very stable helix structure. The double helix structure makes DNA very robust. One ramification of that robustness is that the information stored is reduced by 50% since one strand is totally redundant, being a base-paired complement of the other strand (if one strand is known, so is the other). Although human DNA contains about 12 Gbits of data in its 6 billion bases, only 6 Gbits are information, since the other 6 Gbits in one strand are totally determined by the other strand. Either strand could be considered to be information, but once that choice is made, the other strand adds nothing new, and is therefore not information.

The genetic information system is analogous to a computer's information as determined from information theory, since both are segregated, linear, and digital [Ben73, Cha79]. In bioinformatics (the formal study of the information in life), segregated means each codon is a distinct symbol, linear means these symbols are in a distinct meaningful sequence in the DNA or RNA, and digital means there's no blending of characteristics of symbols and no lowering of fidelity during copying or communicating. As the hype concerning the "Big Switch" to digital TV has indicated – digital is better (it's interesting that life "knew" that from its beginning).

"The genetic information system operates without regard for the significance or meaning of the message, because it must be capable of handling all genetic messages of all organisms, extinct and living, as well as those not yet evolved... The genetic information system is the software of life and, like the symbols in a computer, it is purely symbolic and independent of its environment" [Yoc05p7]. *"Wherever you go in the world, whatever animal, plant, bug or blob you look at, if it is alive, it will use the same dictionary and know the same code. All life is one. The genetic code, bar a few tiny local aberrations, mostly for unexplained reasons in the ciliate protozoa, is the same in every creature. We all use exactly the same language"* [Rid99]. *"It seems that the two-pronged fundamental question: 'Why is the genetic code the way it is and how did it come to be?', that was asked over 50 years ago, at the dawn of molecular biology, might remain pertinent even in another 50 years. Our consolation is that we cannot think of a more fundamental problem in biology"* [Koo08].

The genetic code describes (technically, maps via code bijection) the correspondence between each codon triplet and its corresponding amino acid in a semiotic system that characterizes the symbols and their meaning. A code's symbolic alphabet contains the minimum number of symbols for such mapping. A symbol may have multiple components, such as a codon that has 3 bases, each of which is one-of-four, so the total number of symbols in the genetic alphabet is 64 (= 4^3). The protein alphabet has 20 symbols since there are 20 amino acids.

Since life involves information, it must follow the rules of information theory, including that of transferring information. Yockey proves that it is impossible (zero probability) to transfer information from the 20-symbol protein alphabet to the 64-symbol genetic code. *"Since no code exists to transfer information from protein sequences to mRNA, it is impossible for the origin of life to be 'proteins first'"* [Yoc92]. This categorically eliminates the protein first scenarios as the origin of life, since *"it is mathematically impossible, not just unlikely, for information to be transferred from the protein alphabet to the mRNA alphabet"* [Yoc05p23]. *"Scientists cannot get around it by clever chemistry. This restriction prevails in spite of* [what] *the concentration of protein in a 'prebiotic soup' may have been or may be on some 'Earth-like' planet elsewhere in the universe"* [Yoc05p182]. This is the information theory basis for Crick's "Central Dogma" that states that information transfer is only from DNA to protein, and never the reverse.

Yockey shows that since the genetic code had to be present from the very beginning of life, *"the origin of life, like the origin of the universe*[,] *is unknowable. But once life has appeared, Shannon's Channel Capacity Theorem ... assures us that the genetic messages will not fade away... without the assistance from an Intelligent Designer"* [Yoc05p181]. Although Yockey explicitly distances himself from the ID proponents, many of his findings are in line with the tenets of ID. Yockey has shown the impossibility of transferring information from an alphabet of fewer symbols to one with more basic symbols. As proved in the Appendix C, whatever the source of life (scientifically unknowable), the alphabet involved with the origin of life, by the necessary conditions of information theory, had to be at least as symbolically complex as the current codon alphabet. If intermediate alphabets existed (as some speculate), each predecessor also would be required to be at least as complex as its successor, or Shannon's Channel Capacity [Sha48] would be exceeded for information transfer between the probability space of alphabets with differing Shannon capacity. Therefore, life's original alphabet must have used a coding system at least as symbolically complex as the current codon alphabet. There has been no feasible natural explanation proposed to produce such an alphabet since chance or physicality cannot produce functional information or a coding system, let alone a system as complex as that in life.

Functional information is a useful subset of Shannon information (which is inappropriate to be considered as the only "information" within life, since life not only has complexity, but also functionality). Shannon "information" can be quantified with a functionality variable to calculate Functional bits (Fits) of information. *"This explicitly incorporates empirical knowledge of metabolic function into the measure that is usually important for evaluating sequence complexity"* [Dur07]. Since genes can be thought of as information-processing subroutines, proteins can be analyzed in terms of the products of information interacting with laws of physics to advance our knowledge of the structure and functions of proteins. The patterns of functional information are examined for a protein family. The method proposed is based on mathematical and computational concepts. The resulting equation is $\mathbf{H_f(t) = -\Sigma P(X_f(t)) \log P(X_f(t))}$, where X_f denotes the conditional variable of the given sequence data (X) on the described biological function f, which is an outcome of the variable (F), using the joint variable (X,F). This results in an exponential decrease in probability with a linear increase in functional sequence complexity (FSC). *"For example, 342-residue SecY has a FSC of 688 Fits, but the smaller 240-residue RecA actually has a larger FSC of 832 Fits. The Fit density (Fits/amino acid) is, therefore, lower in SecY than in RecA. This indicates that RecA is likely more functionally complex than SecY"*[Dur07].

A code that defines the meaning of a message symbol or combination of symbols (semiotics) is often used to specify functional information. In a zip code, for example, there is nothing intrinsic in a particular string of 5 digits that causes it to refer to a particular town or portion of a city. It has been established by intelligent beings to have the coded meaning ascribed to it. Those five digits can be decoded during mail processing so the item is delivered to the correct post office. Using zip+4 encoding, delivery can be made to the specific address during decoding.

The American Standard Code for Information Interchange (ASCII) defines the character represented by a particular 7-bit pattern, and is the used for text storage and communication. Imagine the chaos if each manufacturer chose its own arbitrary code to represent characters. What would display if "1100101" were sent to a computer? That particular pattern means 'e' using ASCII, with each other character having its own code. The ASCII standard allows meaningful communication messages to be transferred between equipment programmed to use that standard.

Life has considerable functional information, including multiple coding systems. *"Biosemiotics can be defined as the science of signs in living systems"* [Kul07]. *"Metaphysically disallowing formalism in one's model of reality precludes not only redundancy coding, it precludes semiosis. A purely physical semiotic system cannot exist or function as a messaging system. Representationalism* [symbols represent, e.g. – 'cat' is a representing word] *requires both combinatorial uncertainty* [other letters are possible] *and freedom of deliberate selection... Formalism alone can send and interpret linear digital messages. This remains true even when a material symbol system* [MMS] *with physical symbol vehicles is used by formalism. Polynucleotide genes are such an MSS*" [Abe09G]. It is impossible to send a meaningful message non-symbolically. It has been recognized that *"the cell is a true semiotic system, and that the genetic ... codes are experimental realities"* [Bar08S]. *"Peer-reviewed life-origin literature presupposes that, given enough time, genetic instructions arose via natural events. Thus far, no paper has provided a plausible mechanism for natural-process algorithm-writing... Both the semantics* [meaning] *and syntax* [grammar rules] *of codonic language must translate into appropriate semantics and syntax of protein language. That symbolization must then translate into the 'language' of three-dimensional conformation via minimum-free-energy folding. No combination of the four known forces of physics can account for such conceptual relationships. Symbolism and encryption/decryption are employed. Codons represent functional meaning only when the individual amino acids they prescribe are linked together in a certain order using a different language. Yet the individual amino acids do not directly react physicochemically with each triplet codon. Even after a linear digital sequence is created in a new language, 'meaning' is realized at the destination only upon folding and lock-and-key binding"* [Tre04].

Functional information principles may be used to quantify functional information by calculating a value which *"represents the probability that an arbitrary configuration of a system will achieve a specific function to a specified degree"* [Haz07]. Kalinsky [Kal08] evaluated examples using quantified functional information, comparing the probabilities of such information arising by mindless and designed causes. The five "watermarks" (words whose amino acid letters are meaningful, including "CRAIGVENTER") of Venter Institute's synthetic M. genitalium genome were found to be 10^{22} times more probable to have a designed (known in this case) as opposed to mindless source. A 300 amino acid protein was found to be 10^{155} times more probable to have a designed source. The simplest life form with 382 protein-coding genes [Gla06] was found to be $10^{80,000}$ times more probable to have a designed source. Appendix C has technical details.

Unlike Shannon complexity, which is independent of meaning, functional information assumes the receiver of the information in this book (you, the reader) can appropriately decode the blotches of ink on the white background to gain information based not only on the forms of those blotches, but on knowledge in the receiver's database (you know the meaning of English words). This concept is critical since it is important not only to transmit the messages reliably (a Shannon criterion) from the DNA, but the enzymes and ribosomes must already know how to interpret that coded information in order for proteins (including enzymes), RNA, and replicated DNA to be manufactured appropriately (functional criteria). In other words, the data transmitted may have an effective Shannon information content that exceeds the number of raw data bits transmitted, if one includes all of the information already known by the receiver. When "DNA" appears in this text, for example, considerably more than three English letters is being communicated. If a DNA molecule were placed into a pool of amino acids, the information in the DNA would be useless unless the needed receiver enzymes were present with prior information necessary to function.

Formal algorithmic prescriptive information (PI) for interacting processing systems is unique to life and computer science. Richard Dawkins has noted, *"The machine code of the genes is uncannily computer-like. Apart from differences in jargon, the pages of a molecular biology journal might be interchanged with those of a* *computer engineering journal"* [Daw95p17]. Advances in molecular biology, biochemistry, and bioinformatics have revolutionized our understanding of the cell's miniature world. Cells possess the ability to store, edit, and transmit information and to regulate metabolic and other processes using that information. Whereas cells were once thought of as simple *"homogeneous globules of plasm"* [Hae04] by biologists of Ernst Haeckel's time, *"modern biologists now describe cells as, among other things, 'distributive real-time computers' and complex information processing systems"* [Mey98].

PI is key to any successful computer program (or sub-program), including the programs within life. A successful algorithm will have "computational halting" – that is, the program will stop when (not if) its problem is solved. PI doesn't just describe, it generates meaning and function, providing a recipe or functional algorithm. A programming "bug" is an example of a slight mis-prescription, which makes the program less-than-ideal, or even worse-than-useless (getting no result may be preferable to a wrong result). The essence of prescription itself is choice contingency. Purposeful intent is required at each successive decision node in order to choose configurable switch-settings (or equivalent) and to steer events toward functional results. PI *"instructs or directly produces nontrivial function... Prescriptive information either tells us what choices to make, or it is a recordation of wise choices already made"* [Abe09P]. PI involves formal choices at decision points that cannot be generated by randomness or law (necessity). For example, a computer program is formally solved algorithmically before physically implementing it (a program doesn't just appear on a disk without instantiated choices). The information in life is fundamentally formal, not just physical. Life's information is stored in the physical DNA medium, but the nucleotide sequence of DNA contains PI that is clearly evident.

Cybernetics is the interdisciplinary study of control systems with feedback. Without specifying the origin of cybernetic systems, *"Cybernetics is the study of systems and processes that interact with themselves and produce themselves from themselves"* [Kau07]. It may be tempting to view the physical semiotic systems of life as purely physical. When other cybernetic and artificial life systems are examined, it's clear that they function only because of formal controls instantiated by hardware and software physicality. *"But when it comes to life's syntax, semantics, and pragmatics, we fanatically insist for metaphysical reasons that the system is purely physical. No empirical, rational, or prediction-fulfillment support exists for this dogma"* [Abe09P].

Because of metaphysical beliefs, many scientists recognize only chance and necessity, and dismiss out-of-hand any other non-material reality. This is inconsistent with their accepting the reality of gravity, even though it's unknown why or how it works (but it does -- every time). Note that none of the papers used to establish formalism as a reality invoke anything "supernatural" as its cause. For example, in peer-reviewed papers [Bar08S, Bar08B], Barbieri considers that formal "code-makers" required for biosemiosis may have arisen by purely natural processes. The Salzburg conference on "Natural Genetic Engineering and Natural Genome Editing" investigated how to integrate these formal concepts into understanding evolution [Wit09].

"Biological information is not a substance... biological information is not identical to genes or to DNA (any more than the words on this page are identical to the printers ink visible to the eye of the reader). Information, whether biological or cultural, is not a part of the world of substance" [Hof05]. It is critical to understand the difference between something physical and its symbolic representation. Symbols can be instantiated in any number of different media without changing the meaning of those symbols. For example, a word can appear in printed, hand-written, projected, electronically transmitted, encrypted, etc. form without changing its meaning.

"Genetic algorithms instruct sophisticated biological organization. Three qualitative kinds of sequence complexity exist: random (RSC), ordered (OSC), and functional (FSC). FSC alone provides algorithmic instruction... Law-like cause-and-effect determinism produces highly compressible order. Such forced ordering precludes both information retention and freedom of selection so critical to algorithmic programming and control. Functional Sequence Complexity requires this added programming dimension of uncoerced selection at successive decision nodes in the string. Shannon information theory measures the relative degrees of RSC and OSC... [and] *cannot measure FSC. FSC is invariably associated with all forms of complex biofunction, including biochemical pathways, cycles, positive and negative feedback regulation, and homeostatic metabolism. The algorithmic programming of FSC, not merely its aperiodicity, accounts for biological organization. No empirical evidence exists of either RSC or OSC ever having produced a single instance of sophisticated biological organization. Organization invariably manifests FSC rather than successive random events (RSC) or low-informational self-ordering phenomena (OSC)"* [Abe05]. For example, a fractal produces amazingly intricate visual patterns, but is an example of OSC whose recursive algorithm contains little information.

The rise of PI presumably occurred early in the evolutionary history of life which had several organisms that depended upon nearly 3,000 highly coordinated genes in cybernetic systems. Genes are linear, digital, cybernetic sequences. They are meaningful, pragmatic (functional), physically instantiated recipes. Bioinformation has been selected algorithmically at the sequence level to instruct eventual three-dimensional shape. The shape is specific for a certain structural, catalytic, or regulatory function, with all functions integrated into a symphony of metabolic functions. The sequence of DNA nucleotides ultimately produces the protein's shape, but using a very indirect mechanism that includes multiple conversions of digital information. It has long been recognized that the genotype is a generative algorithm – *"a carefully spelled out and foolproof recipe for producing a living organism... the algorithm must be written in some abstract language... a language must have rules"* [Ede66].

Constraints and controls must be differentiated. Investigators can choose initial conditions for the starting point of their experimentation, but those "chosen constraints" can be considered controls because they were deliberately selected to steer events toward desired results. Choice contingency alone, not the natural law constraints themselves, achieves non trivial integration, organization, and function. *"Controls do not arise from the categories of chance contingency and necessity addressed by thermodynamics, kinetics and physics in general. Physics can address constraints. Physics cannot address bona fide controls without acknowledging the reality of non naturalistic engineering. Life is wholly dependent upon tight regulation and controls. For this reason, physics and chemistry alone cannot adequately address and explain life any more than physics and chemistry alone can explain engineering"* [Abe10]. Formal choice contingency can control physicality.

"What kind of information produces function? In computer science, we call it a 'program.' ... No man-made program comes close to the technical brilliance of even Mycoplasmal genetic algorithms. Mycoplasmas are the simplest known organism with the smallest known genome, to date. How was its genome and other living organisms' genomes programmed?" [Abe05] This is probably the biggest unanswered question in biology since algorithms are unknown except for choice contingency. Both computer computations and the end-products of biochemical pathways require non-random algorithmic processes or procedures that produce the needed results Those algorithms are never "self-ordered" by redundant cause-and-effect necessity. Each successive nucleotide in DNA or RNA is a quaternary (base-4) "switch setting," produced by uncoerced selection. This cybernetic aspect of life processes is directly analogous to that of other computer programming, and requires considerably more emphasis and attention than it typically receives. *"What sense can we make, then, of the PI found in*

nature and particularly in any theorized primordial biosemiosis? Random coursing through a succession of bifurcation [decision forking] *points has never been observed to lead to prescription of function, computational halting, sophisticated circuitry, or system organization. The self-ordering events described by chaos theory cannot generate conceptual formal organization. Semiosis, cybernetics, and formal organization all require deliberate programming decisions, not just self-ordering physicodynamic redundancy. Self-ordering phenomena are low-informational, highly redundant, unimaginative, and usually destructive of organization (e.g., tornadoes and hurricanes). No prediction fulfillments have been realized of spontaneous natural events producing formal algorithmic optimization. No empirical support or rational plausibility exists for blindly believing in a relentless natural-process assent up the foothills of a rugged fitness landscape toward mountain peaks of formal functionality. Investigator involvement creates this illusion usually through the hidden artificial steering of experimental iterations"* [Abe09P].

Life is orchestrated by the prescriptive information (instruction) content and the inherent systems it has from its inherited genome. Cells also are dependent upon their current environment, especially for energy, whose transducing mechanisms are instructed by its genome. Life is an integration of many algorithmic processes that give rise to biofunction and overall life-maintaining metabolism. Each symbolic choice is critical to the determination of eventual function, so that bioinformation is more than just physical order, complexity, or probabilistic uncertainty, but is an instructive, orchestrational recipe. Unlike Shannon complexity, PI is concerned with functionality, not the degree of compressibility.

Take note anyone that wants to earn an "easy" $1 million: *"The Origin-of-Life Prize® ... will be awarded for proposing a highly plausible mechanism for the spontaneous rise of genetic instructions in nature sufficient to give rise to life"* [OOLprize]. *"The Origin-of-Life Foundation, Inc. is a science and education foundation encouraging the pursuit of natural-process explanations and mechanisms within nature."* Since no theory of genetic information is complete without a model of mechanism for the source of such prescriptive information within Nature, *"all submissions must address the source of the prescriptive information through non-supernaturalistic natural processes. Which of the four known forces of physics, or what combination of these forces, produced prescriptive, functional information, and how? What is the empirical evidence for this kind of prescriptive information (instruction) spontaneously arising within Nature?"* [OOLprize].

In all known life, genomes manifest functional coded messages, using a sign system, to distant sites through an information channel to energy-consuming decoding receivers in ribosomes. Languages are translated from one symbolic, indirect representation of the message from one alphabet into another, specifically from the codon language into the language of the physical amino acid sequence in the end-product. Explanation is required for the instructions capable of causing and affecting the multitude of individual manufacturing processes and coordinating of all of those diverse processes toward the apparent "purpose" of being alive. What natural mechanism(s) caused *"the initial writing of this prescriptive information by nature, not just the modification of existing genetic instruction through mutation"* [OOLprize].

The Genetic Selection (GS) Principle differentiates after-the-fact modification of an existing genome from the production of the genome from non-living components. The GS Principle *"distinguishes selection of existing function (natural selection) from selection for potential function (formal selection at decision nodes, logic gates and configurable switch-settings)"* [Abe09G]. Undirected nature has no "purpose," and the source of functional algorithmic PI that "will have purpose" at some future time has yet to be explained. If RNA-first is true, *"the real issue of life origin lies in answering how the initial single positive strands of RNA instructions got sequenced so as to prescribe microRNA regulation, amino acid sequencing and eventual folding function. No new information is generated in base-pairing replications...* [which is] *quite secondary to the already-programmed, formal, linear digital instructions of the single positive strand... Prior to an algorithm having computational function, no basis exists for selection in nature... How did any computational program arise in nature? Computation is formal, not physical. Natural selection cannot generate formalisms. It can only prefer the results of formal computations—already living organisms... No organism exists without hundreds of cooperating formal algorithms all organized into one holistic scheme. The more computational steps that are required to achieve integrated halting, the harder it becomes for an inanimate environment to explain optimization"* [Abe09G].

"How did inanimate nature give rise to an algorithmically organized, semiotic and cybernetic life? Both the practice of physics and life itself require traversing not only an epistemic cut, but a Cybernetic Cut – a fundamental dichotomy of reality" [Abe08]. The "epistemic cut" describes the unavoidable gulf between an object and knowledge about that physical object (between description and the thing being described) [Pat67]. *"The Cybernetic Cut must be crossed to program computational halting into any form of physical hardware. To prescribe, instruct or program formal*

utility is to traverse The Cybernetic Cut... The bifurcation points found in the simplest binary system of choice contingency are bona fide decision nodes. Crossing the Cybernetic Cut requires the ability to purposefully steer through successive bifurcation points down a path toward a desired goal... Bifurcation points, in the absence of the intentional choice that would convert them to true decision nodes, consistently fail to generate sophisticated utility. In symbol systems, the randomization of symbols and denial of intentional symbol selection quickly leads to the loss of even rudimentary meaning and function... What exactly is the missing ingredient that renders life unique from inanimate physics and chemistry? The answer lies in the fact that life, unlike inanimacy, crosses the Cybernetic Cut" [Abe08]. There is *"a critically important distinction between order and the informed functional organization that characterizes living systems... our paradigm is cybernetic"* [Cor00]. True organization requires crossing the Cybernetic Cut in order to produce a cybernetic system. The resulting system is functional using physical components, but physicality has no mechanism to produce a cybernetic system requiring formal organization.

The RNA/DNA sequence structure is formed prior to protein folding of the PI's result. The fixed linear digital highly informational nucleotide sequence must be prescribed semiotically before the weak hydrogen-bond folding of the secondarily protein structure occurs. The Genetic Selection Principle, *"states that selection must operate at the genetic level, not just at the phenotypic level, to explain the origin of genetic prescription of structural and regulatory biological function. This is the level of configurable switch-settings (nucleotide selection). Selection must first occur at each decision node in the syntactical string. Initial programming function cannot be achieved by chance plus after-the-fact selection of the already-existing fittest programs (phenotypes). Evolution is nothing more than differential survival and reproduction of already-existing fittest phenotypes. The computational programming proficiency that produced each and every phenotype must first be explained... Thus far, no natural process explanation has been published for selection at the decision-node, configurable-switch, nucleotide selection level"* [Abe08]. No feasible source of life's prescriptive algorithms has been proposed to date. The speculations proposed thus far are"dreams," based on the belief that physicality is the only reality, and therefore "must have" produced life's cybernetic complexity. Information science, by contrast, recognizes non-material choice contingency as the only reality capable of producing PI.

Computer/electronic engineers have designed the hardware and interpreting microcode to read the computer's native language instructions (non-physical machine language) and execute the operations specified by each instruction. A higher-level computer language (such as BASIC, C, or FORTRAN) may be used to translate a desired algorithm into the computer's native language. A computer's operating system (OS), such as Linux or Windows, is a set of programs that allows applications programs to execute on the computer's hardware, allowing access to storage and input/output devices. To the computer hardware, an OS is just another program, unless it's built into the hardware. To an application program, typically only the OS (not the hardware) is visible. In the history of computers, there has never been an instance of functional hardware or software arising by undirected processes (although some non-functional student-written programs have had the appearance of being the result of random trials).

Advances in molecular biology, biochemistry, and bioinformatics (the study of the information of life) have revolutionized our understanding of the cell's miniature world. Cells possess the ability to store, edit, and transmit information and to regulate metabolic and other processes using that information. From the information perspective, the genetic system is a preexisting operating system of unknown origin that supports the storage and execution of a wide variety of specific genetic programs (the genome applications), each program being stored in DNA. DNA is a storage medium, not a computer, that specifies all information needed to support the growth, metabolism, parts manufacturing, etc. for a specific organism via gene subprograms. DNA has been compared to a computer's disk drive [D'On10], which makes sense in a NUMA (non-uniform memory access) model. Early real computers used disk-like drums and other sequential-access main memories. This author has peer-reviewed publications [Joh95, Joh97B, Joh05] describing concepts of distributed sequentially-accessed special-purpose and general memories (analogous to those in life) in heterogeneous (different) multiprocessor systems. In cells, there are many RNAs and micro-proteins which may function as registers and inter-processor communications channels.

Biologically-inspired computer systems began in 1994 [Adl94], with potential Turing-completeness later proven [Bon96]. Enzyme/DNA computers were proposed in 2002 [Lov03] and implemented in 2004 [Ben04]. A distributed computing system has been designed, patterned after life's computing systems [Bab06], and a massively parallel computing system was designed based on biological cells [Ban10]. Computer hardware [D'On10] and operating systems [Yan10] comparisons have also been made with life's processing systems to find functional equivalences between the engineered and biological systems. Eventually, computer engineers may be able to harness the tremendous processing power exhibited in life.

Technically, DNA is an example of shared memory in a distributed heterogeneous multiprocessor system with Flynn classification multiple input streams and multiple output streams [Fly72]. Each cell has over 2,000 different enzyme computers that simultaneously read different portions of the DNA genetic code, each producing its own output (for example, via mRNA). The information is processed according to the individual programs, many operating independently (though many operations require multiple cooperating enzymes), often using coding systems (e.g. – codon-based encryption). Often, smaller sections of RNA from different genes are spliced together to form the mRNA for a particular protein specification. The mRNA output is ultimately to another OS in a ribosome, which has its own program stored in its RNA, where the codes are decrypted. The needed output signals are then transmitted to the tRNA computer (which has been programmed to pick up its associated amino acid via its own program and OS) so that the amino acid specified by the codon is transported to the construction site to be added to the protein being built.

Every case of coded information, where the source is known, invariably requires mind for its creation. *"Neither order nor complexity is the determinant of algorithmic function... This is one of most poorly understood realities in life-origin science. Selection alone produces functionality. Without selection, evolution would be impossible... A 'cybernetic program' presupposes a cybernetic context in which it operates. One has to have an operating system of 'rules' before one can have an application software. And of course one must have a hardware system too. All of these components only come into existence through 'choice contingency,' not through 'chance contingency' or law. One of many problems with metaphysical materialism is that it acknowledges only two subsets of reality: chance and necessity. Neither can write operating system rules or application software. Neither can generate hardware or any other kind of sophisticated machinery, including molecular machines (the most sophisticated machinery known)"* [Abe07].

"The reductionist approach has been to regard information as arising out of matter and energy. Coded information systems such as DNA are regarded as accidental in terms of the origin of life and that these then led to the evolution of all life forms as a process of increasing complexity by natural selection operating on mutations on these first forms of life" [McI09]. In the last 10 years, at least 20 different information codes were discovered in life, each operating according to arbitrary conventions (not law).

Examples include RNA codes [Far07], metabolic codes [Bru07], cytoskeleton codes [Gim08], histone codes [Jen01], and alternative splicing codes [Bar10].

Much of the functional information of life is involved with cybernetic control, in which life's components regulate and control where needed. *"All the equations of physics taken together cannot describe, much less explain, living systems. Indeed, the laws of physics do not even contain any hints regarding cybernetic processes or feedback control. Thus, the term 'dissipative structures' does not adequately describe the 'informed', purposive organization of living systems. It is comparable to characterizing jet engines -- which are painstakingly designed and manufactured with extremely precise dimensional properties and tolerances -- as dissipative structures. They are neither self-designed nor are their dissipative properties among their most salient features"* [Cor05]. Many RNAs are transcribed from DNA to be regulators (necessary for "control"), including of mRNA. *"In addition to revealing the surprising diversity of post-transcriptional events that regulate mRNAs, our work also points to new roles for a family of proteins that mediate RNA interference"* [Han10]. It has been discovered that *"proteins communicate not by a series of simple one-to-one communications, but by a complex network of chemical messages"* [Edi10].

Researchers have been able to identify a mechanism that turns a maturing embryo's developmental genes off and on. *"By understanding how development unfolds, we can better control this process, which includes cell proliferation and organ development"* [Yeh10]. Scientists have found that many proteins are multi-functional from the transcription, metabolism, and manufacturing levels. *"At all three levels, we found M. pneumoniae was more complex than we expected"* [Ser09]. Proteins in the NPC (nuclear pore complex) are not only communication channels that regulate the passage of all molecules (e.g. mRNA) to and from a cell's nucleus, but also *"play a role in the organization of the genome and a very direct role in gene expression"* [Sal10] (genes can be turned on or off). The NPC probably also does proof-reading on mRNA while passing through, since it "stalls" at the interface [Gru10].

"The complexity of biology has seemed to grow by orders of magnitude... the signaling information in cells is organized through networks of information rather than simple discrete pathways" [Hay10]. Biocomplexity professor Stuart Kauffman points out, *"the genetic regulatory network in humans has some 23,000 genes, among which are at least 2,040 transcription factor genes. These TFs regulate one another's transcriptional activity and those of genes that are regulated but not regulating. Work*

on yeast gene networks shows that they appear to be one large interconnected network... This genetic regulatory network is a non-linear dynamical system of high complexity. Modeling genes as binary, on, off, devices and studying large 'random Boolean networks' has shown that these networks, and piecewise linear networks, and linear ordinary differential equation networks all show the same generic behaviors" [Maz10p223], all being complex data processing implementations.

Scientists are investigating *"the organization of information in genomes and the functional roles that non-protein coding RNAs play in the life of the cell. The most significant challenges can be summarized by two points: a) each cell makes hundreds of thousands of different RNAs and a large percent of these are cleaved into shorter functional RNAs demonstrating that each region of the genome is likely to be multifunctional and b) the identification of the functional regions of a genome is difficult because not only are there many of them but because the functional RNAs can be created by taking sequences that are not near each other in the genome and joining them together in an RNA molecule. The order of these sequences that are joined together need not be sequential. The central mystery is what controls the temporal and coordinated expression of these RNAs"* [Gin10]. *"It is very difficult to wrap your head around how big the genome is and how complicated... It's very confusing and intimidating... The coding parts of genes come in pieces, like beads on a string, and by splicing out different beads, or exons, after RNA copies are made, a single gene can potentially code for tens of thousands of different proteins, although the average is about five... It's the way in which genes are switched on and off, though, that has turned out to be really mind-boggling, with layer after layer of complexity emerging"* [LeP10].

For functional communication (including controls) to occur, both sender and receiver of each communication step must know the communication protocol and how to handle the message. In each cell, there are multiple operating systems, multiple programming languages, encoding/decoding hardware and software, specialized communications systems, error detection and correction mechanisms, specialized input/output channels for organelle control and feedback, and a variety of specialized "devices" to accomplish the tasks of life. The author can attest that these concepts are not trivial since many were fundamental to his second Ph.D. thesis research [Joh97T].

The challenge for an undirected origin of such a cybernetic complex interacting computer system is the need to demonstrate that the rules, laws, and theories that govern electronic computing systems and information don't apply to the even more complex digital information systems that are in living organisms. Laws of chemistry and physics, which follow exact statistical, thermodynamic, and spacial laws, are totally inadequate for generating complex functional information or those systems that process that information using prescriptive algorithmic information. Unfortunately, most people investigating origins are unfamiliar with the immensity of the problems, and believe that time, chance, and natural selection can accomplish almost anything.

It can correctly be asserted that based on known science, some things are impossible. Based on the second law of thermodynamics (entropy/disorganization always increases), a perpetual motion machine is impossible. One needn't prove that each machine that could be conceived is impossible. If it is asserted that such a machine is possible, the one making such an assertion would need to first show that known science is wrong concerning increasing entropy before the assertion is given consideration as science. All the known laws, theorems, and principles of information science indicate that codes, complex functional information, and prescriptive algorithmic information cannot arise from physicality. Therefore, based on currently accepted information science, but realizing that science is always subject to change, it seems impossible for life to have arisen purely from physicality. Those holding the view that physicality can cause functional information would need to falsify the nine null hypotheses of Appendix D before science recognizes any scientific validity of that view.

Those who insist on purely physical causes of life are thus in an untenable position when it comes to known science. Not only can they not prove that it's possible (non-zero probability) for life to come about by the physical interactions of nature, but the information content of life precludes that possibility. *"There is no known law of nature, no known process*[,] *and no known sequence of events which can cause information to originate by itself in matter*" [Git97p107].

Note that one cannot use the information of life to "prove" that information can arise purely from physicality, as that would simply be a tautology based on the assumption of life from physicality, with no direction. In examining any complex functional information where the source of the information is known, it invariably (no known exceptions) resulted from action(s) of intelligent agent(s), either directly or indirectly (such as through a computer program).

6 Increasing Complexity of Life

"We have seen that living things are too improbable and too beautifully 'designed' to have come into existence by chance. How, then, did they come into existence? The answer, Darwin's answer, is by gradual, step-by-step transformations from simple beginnings, from primordial entities sufficiently simple to have come into existence by chance. Each successive change in the gradual evolutionary process was simple enough, relative to its predecessor, to have arisen by chance. But the whole sequence of cumulative steps constitutes anything but a chance process, when you consider the complexity of the final end-product relative to the original starting point. The cumulative process is directed by nonrandom survival" [Daw96Bp43]. Darwin wrote *"If it could be demonstrated that any complex organ existed which could not possibly have been formed by numerous, successive, slight modifications, my theory would absolutely break down"* [Dar98p154]. *"The Darwinian theory is in principle capable of explaining life. No other theory that has ever been suggested is in principle capable of explaining life"* [Daw96Bp288]. *"The following problems have proven utterly intractable not only for the mutation-selection mechanism but also for any other undirected natural processes proposed to date: the origin of life, the origin of the genetic code, the origin of multicellular life, the origin of sexuality, the scarcity of transitional forms in the fossil record, the biological big bang that occurred in the Cambrian era, the development of complex organ systems, and the development of irreducibly complex molecular machines"* [Dem99p29].

Evolution is this chapter's topic: *"Biological evolution, simply put, is descent with modification. This definition encompasses small-scale evolution (changes in gene frequency in a population from one generation to the next) and large-scale evolution (the descent of different species from a common ancestor over many generations)"* [EVsite]. Microevolution, small adaptive changes that are heritable, is accepted as verifiable fact by virtually all scientists. Macroevolution, the scenario that all life originated by undirected natural processes from an original organism, is believed by most scientists, but doubted by many scientists, including over 800 who are signers of: *"We are skeptical of claims for the ability of random mutation and natural selection to account for the complexity of life. Careful examination of the evidence for Darwinian theory should be encouraged"* [Dis-web]. Over 300 physicians have signed: *"As medical doctors we are skeptical of the claims for the ability of random mutation and natural selection to account for the origination and complexity of life and we therefore* *dissent from Darwinian macroevolution as a viable theory. This does not imply the endorsement of any alternative theory"* [PSSI]. *"A persistent debate in evolutionary biology is one over the continuity of microevolution and macroevolution – whether macroevolutionary trends are governed by the principles of microevolution"* [Sim02].

Many extend evolution back to the prebiotic era, so that abiogenesis (covered in chapter 4) is also included in evolution. This chapter will examine what is known about DNA changes through time, development of new morphology, irreducible complexity, the Cambrian explosion, the biological tree, what increasing information would entail, computer simulations, and genetic algorithms. A major focus will be examining beliefs that are purported to be science concerning undirected evolution to ascertain their scientific validity. The scientific validity of intelligent design views will also be examined, including that of ID as a superset of evolution. Keep in mind that the vast majority of science is currently directed toward supporting Darwinism, and this chapter presents "the minority" scientific evidence, which hopefully will encourage a more critical analysis of all findings (how strongly do data support conclusions?).

Only genetic changes can affect evolution, as those changes can be inherited by offspring. Mutations in DNA are normally considered random, and usually occur during cell replication. Approximately one in 100,000 copied base-pairs has an error [Dar86], but the proof-reading reduces point mutation error rate to between 10^{-11} to 10^{-9}. Mutations of genome segments can include duplication, inversion, transposition, insertion, or deletion [Spe97p40]. After mutation, a gene is duplicated (usually the mutation prevents duplication) with the same fidelity as any gene. If a codon is mutated, it may create a different amino acid in the protein it encodes, unless it mutates to a redundant codon for the same amino acid (a built-in error tolerance).

Neo-Darwinian evolution is described by Dawkins [Daw96C] as random (chance) mutation followed by selection, with selection favoring no mutations. *"Mutation is not an increase in true information content, rather the reverse* [an information loss], *for mutation, in the Shannon analogy, contributes to increasing the prior uncertainty. But now we come to natural selection, which reduces the 'prior uncertainty' and therefore, in Shannon's sense, contributes information to the gene pool. In every generation, natural selection removes the less successful genes from the gene pool, so the remaining gene pool is a narrower subset... what is the information about? It is about how to survive"* [Daw08I]. Note that the first organism and asexual reproducers have no "gene pool" from which to draw. Neo-Darwinism breaks down the improbability of macroevolution into small changes that allow scaling up the backside of *"Mount Improbable ... inch by million-year inch"* [Daw96Cp77]. In this view, the changes can accumulate in the DNA until a positive gene is formed (at that time, only genes that had physical manifestations were thought to contribute to a selective advantage), at which point that gene will be selected for continuation in the new organism's DNA. The non-coding part of DNA became known as "Junk DNA" which is *"the remains of nature's experiments which failed"* [Ohn72].

Dawkins popularized the idea that any DNA not actively trying to get to the next generation would slowly decay away through mutation and that genes are the basis of evolutionary selection [Daw76]. Sagan writes concerning junk DNA *"some, maybe even most, of the genetic instructions must be redundancies, stutters, and untranscribable nonsense. Again we glimpse deep imperfections at the heart of life"* [Sag92]. Non-coding sections of DNA are seen as the result of mutations that haven't yet resulted in formation of useful genes so that they would provide a selective advantage. This theme was echoed in authoritative textbooks also. *"Introns have accumulated mutations rapidly during evolution, and it is often possible to alter most of an intron's nucleotide sequence without greatly affecting gene function. This has led to the suggestion that intron sequences have no function at all and are largely genetic 'junk'"* [Alb94]. *"Much repetitive DNA serves no useful purpose whatever for its host. Rather, it is selfish or junk DNA, a molecular parasite that, over many generations, has disseminated itself throughout the genome"* [Voe95]. As late as 2004, Dawkins wrote *"there's lots more DNA that doesn't even deserve the name pseudogene. It, too, is derived by duplication, but not duplication of functional genes. It consists of multiple copies of junk, 'tandem repeats', and other nonsense which may be useful for forensic detectives but which doesn't seem to be used in the body itself. Once again, creationists might spend some earnest time speculating on why the Creator should bother to litter genomes with untranslated pseudogenes and junk tandem repeat DNA"* [Daw98, Daw04]. Biology professor PZ Myers wrote in 2010, *"the genome is mostly dead, transcriptionally. The junk is still junk"* [Mye10].

While the mutation mechanism for gene formation makes an interesting story, there are a number of scientific difficulties with the scenario. Blind chance is the only known mechanism possible for such gene formation since a selective advantage was thought to require some manifestation that is genetically coded (such as making a particular protein). Since nothing prevents what would have become a mutated "correct" nucleotide from mutating again to become useless (mutation is by chance), the probability for a useful mutated gene is that all required mutations take place in one organism before or during reproduction. A typical gene contains over 1,000 base-pairs, but if we speculate that an operative gene could have only the codons required for encoding a 50-amino acid protein (ignoring introns) the minimum length of the gene would be at least 150 base-pairs. Since the Shannon information of such a hypothetical gene is unknown, the probability of forming such a gene can be estimated from $4^{-150} = 5 \times 10^{-91}$ (that's for one specific gene in one organism, which may decrease for larger gene size or increase if additional functional genes or codon aliases are considered). Since "useless" base-pairs have no advantage and since transferring them would take additional time and require at least 0.035 electron volts per bit for each step during replication [Yoc05p25] (there's no free lunch!), any mutation that eliminated those useless base-pairs would have a selective advantage so it would make sense that they would be removed from the genome long before they had a chance to form something with a selective advantage. Time is actually an enemy since the 2nd law will tend to randomize genetic content unless directed energy prevents that from happening (there would be no advantage in directing energy to "useless" structures).

"Junk DNA" has been classified as a misnomer by ID proponents as early as 1986 [Den86], since *"Junk DNA and directed evolution are in the end incompatible concepts"* [Den98]. The journal Science refused to print a 1994 letter pro-ID scientist Forrest Mims wrote warning about assuming that "junk DNA" was useless [Mim94]. Studies have shown that non-coding DNA plays a role in embryonic development in such areas as the reproductive tract [Kep96] and the central nervous system [Koh96]. *"Geneticists have long focused on just the small part of DNA that contains blueprints for proteins. The remainder--in humans, 98 percent of the DNA–was often dismissed as junk. But the discovery of many hidden genes that work through RNA, rather than protein, has overturned that assumption. These RNA-only genes tend to be short and difficult to identify. But some of them play major roles in the health and development of plants and animals"* [Gib03]. *"Some scientists now suspect that much of what makes one person, and one species, different from the next are variations in the gems hidden within our 'junk' DNA"* [Gib03]. At first, evolutionists thought that introns had no role in the production of proteins and regarded them as merely junk. However, research has proven that they play vitally important roles. *"For years, more and more research has, in fact, suggested that introns are not junk but influence how genes work... introns do* *have active roles"* [Ray03]. *"Just when scientists thought they had DNA almost figured out, they are discovering in chromosomes two vast, but largely hidden, layers of information that affect inheritance, development, and disease"* [Gib03]. Genetic biologist Wojciech Makalowski states: *"Now, more and more biologists regard repetitive elements as a genomic treasure... that repetitive elements are not useless junk DNA but rather are important, integral components of eukaryotic genomes"* [Mak03]. *"Nonprotein coding RNA (ncRNA) refers to mRNA that is transcribed from DNA but not translated into protein. Rather than being 'junk' DNA (ie an evolutionary relic) some nonprotein coding transcripts may in fact play a critical role in regulating gene expression and so organizing the development and maintenance of complex life"* [Per05]. *"Scientists are puzzling over a collection of mystery DNA segments that seem to be essential to the survival of virtually all vertebrates. But their function is completely unknown. The segments, dubbed 'ultraconserved elements', lie in the large parts of the genome that do not code for any protein. Their presence adds to growing evidence that the importance of these areas, often dismissed as junk DNA, could be much more fundamental than anyone suspected"* [Pea04].

"If one adds together nucleotides that are individually nonfunctional, one may end up with a sum of nucleotides that are collectively functional. Nucleotides belonging to chromatin are an example. Despite all arguments made in the past in favor of considering heterochromatin as junk, many people active in the field no longer doubt that it plays functional roles... Nucleotides may individually be junk, and collectively, gold" [Zuc97]. John Mattick, Director of the Centre for Molecular Biology and Biotechnology at the University of Queensland, writes: *"I think this will come to be a classic story of orthodoxy derailing objective analysis of the facts, in this case for a quarter of a century, the failure to recognize the full implications of this — particularly the possibility that the intervening noncoding sequences may be transmitting parallel information in the form of RNA molecules—may well go down as one of the biggest mistakes in the history of molecular biology... Indeed, what was damned as junk because it was not understood may, in fact, turn out to be the very basis of human complexity"* [Mat03]. Realize that since every protein is the result of the execution of a prescriptive algorithm, what had been assumed is that random memory changes can produce a new functional program, a concept foreign to information science.

Researchers are discovering that what had been dismissed as evolution's relics are actually vital to life. What used to be considered evidence for neo-Darwinism gene-formation mechanism can no longer be used as such evidence. In this case, neo-Darwinism has been a proven science inhibitor as it postponed serious investigation of the non-coding DNA within the genome. This is reminiscent of the classification of 86 (later expanded to 180) human organs as "vestigial" that Robert Wiedersheim (1893) believed *"lost their original physiological significance,"* in that they were vestiges of evolution [Wie1893]. Functions have since been discovered for all 180 organs that were thought to be vestigial, including the wings of flightless birds, the appendix, and the ear muscles of humans [Ber90].

It should also be noted that just having a selective advantage doesn't guarantee that such an advantage will continue to be inherited. The selective value (S) of a phenotype can express the relative increase (or decrease, if negative) in reproductive probability, such that a mutated organism with S of 0.1% would have 0.1% more offspring than the average organism. The probability of that mutant strain surviving, assuming equal mating probability, is $2S/(1 - e^{-2SN})$ [Spe97p80], where N is the total population size. Even with S of 0.1%, in a population of 1 million, the probability of survival of that mutant strain is only about 10^{-6}, whereas in a population of 100, the probability of survival increases to 0.011 (which has led many to believe macroevolution only occurs in isolated groups of organisms in what is known as "punctuated equilibria" [Eld72]). In other words, just having a selective advantage does not mean the mutant will survive. Another problem is the cost of substitution that can occur in a given time: *"natural selection cannot occur with great intensity for a number of characters at once unless they happen to be controlled by the same genes"* [Hal57]. Haldane defines intensity of selection for a genotype survival to reproductive age (s_0) versus the entire population (S) as $I = \log_e(s_0/S)$. *"It takes as many deaths, or their equivalents, to replace a gene by one producing a barely distinguishable phenotype as by one producing a very different one. If two species differ at 1000 loci, and the mean rate of gene substitution, as has been suggested, is one per 300 generations, it will take 300,000 generations to generate an interspecific difference"* [Hal57p521]. The replacement of a phenotype with a more desirable one would take many generations as the favorable genes spread in the population by the lowered fertility (or death) of those without those genes.

Functional information increasing as one traverses the tree upward from its root is another inexplicable feature of Darwinism. *"The pressure to code information in a fault-tolerant manner implies that codes should evolve that are robust to deleterious* *mutations*" [Ofr03]. There is no doubt that mutations can modify existing information within the genome to produce modified characteristics. Segment and point mutations can cause a loss of information, and that loss may actually increase survivability. For example, if a bacterium mutates at the point where an antibiotic would attach, it becomes resistant to that antibiotic. The "best" place to become infected with such bacteria is a hospital, because nearly all of the "normal" bacteria have been killed, leaving only the "deformed" bacteria to compete for survival. Sickle cell anemia is caused by a point mutation in the hemoglobin beta gene on chromosome 11. While normally this defect is detrimental, the plasmodium parasite that causes malaria has difficulty invading cells with that mutation, so those having this genetic defect are protected against malaria (a selective advantage in high malaria regions). The nylon-digesting bacterium [Kin75] is often cited as an example of information increase via mutation, especially since nylon didn't exist before 1935. This may be the result of a "frame-shift" (with decoding off by one nucleotide) mutation that enables the bacterium to digest nylon at about 8% efficiency. Since it can no longer digest its normal diet of cellulose, a net functional information loss is evident in the genome.

"There is no evidence that genetic information can build up through a series of small steps of microevolution... Mutations reduce the information in the gene by making a protein less specific. They add no information, and they add no new molecular capability... None of them can serve as an example of a mutation that can lead to the large changes of macroevolution... The failure to observe even one mutation that adds information is more than just a failure to find support for the theory. It is evidence against the ... neo-Darwinian theory" [Spe97p159-160]. Evolutionary geneticist H. Orr writes supporting punctuated equilibrium in some cases: *"We conclude – unexpectedly – that there is little evidence for the neo-Darwinian view: its theoretical foundations and the experimental evidence supporting it are weak"* [Orr92p726]. *"Stunningly, information has been shown not to increase in the coding regions of DNA with evolution. Mutations do not produce increased information... the amount of coding in DNA actually decreases with evolution... No increase in Shannon or Prescriptive information occurs in duplication"* [Abe09G]. *"None of the papers published in JME (Journal of Molecular Evolution) over the entire course of its life (1971-) as a journal has ever proposed a detailed model by which a complex biochemical system might have been produced in a gradual, step-by-step Darwinian fashion"* [Beh96p176].

"We must concede there are presently no detailed Darwinian accounts of the

evolution of any biochemical or cellular system, only a variety of wishful speculations" [Har01]. Biologist Lynn Margulis writes: *"We agree that very few potential offspring ever survive to reproduce and that populations do change through time, and that therefore natural selection is of critical importance to the evolutionary process. But this Darwinian claim to explain all of evolution is a popular half-truth whose lack of explicative power is compensated for only by the religious ferocity of its rhetoric. Although random mutations influenced the course of evolution, their influence was mainly by loss, alteration, and refinement. One mutation confers resistance to malaria but also makes happy blood cells into the deficient oxygen carriers of sickle cell anemics. Another converts a gorgeous newborn into a cystic fibrosis patient or a victim of early onset diabetes. One mutation causes a flighty red-eyed fruit fly to fail to take wing. Never, however, did that one mutation make a wing, a fruit, a woody stem, or a claw appear. Mutations, in summary, tend to induce sickness, death, or deficiencies. No evidence in the vast literature of heredity changes shows unambiguous evidence that random mutation itself, even with geographical isolation of populations, leads to speciation. Then how do new species come into being?"* [Mar03] *"Yes, small-scale evolution is a fact, but there is no reason to think it is unbounded. In fact, all our data suggests that small-scale evolution cannot produce the sort of large-scale change Darwinism requires"* [Hun03]. Anthropologist Roger Lewin, at the Conference on Macroevolution, said: *"The central question of the Chicago conference was whether the mechanisms underlying microevolution can be extrapolated to explain the phenomena of macroevolution. At the risk of doing violence to the position of some people at the meeting, the answer can be given as a clear, No"* [Lew80].

The fossil record is believed by many to be a support for Darwinism, but there are many problems therein. The Cambrian explosion refers to the geologically sudden (no more than 0.1% of earth's history) appearance of at least 19 (as many as 35 of the 40 total) phyla of animals in the fossil record during the Cambrian period. During this event, each phylum exhibits a unique architecture, blueprint, or structural body plan with no predecessors or intermediaries. Examples of basic animal body plans include cnidarians (corals and jellyfish), mollusks (squids and shellfish), arthropods (insects, crustaceans, and trilobites), and chordates (vertebrates, including humans). *"Most of the animal phyla that are represented in the fossil record first appear, 'fully formed,' in the Cambrian some 550 million years ago... The fossil record is therefore of no help with respect to the origin and early diversification of the various animal phyla"*

[Bar01]. *"If subphyla are included in the count of animal body plans, then at least thirty-two and possibly as many as forty-eight of fifty-six total body plans (57.1 to 85.7 percent) first appear on earth during the Cambrian explosion"* [Mey03p330]. *"All living phyla may have originated by the end of the explosion"* [Val99p327].

Carl Woese, microbiologist and originator of the RNA world hypothesis, writes: *"Phylogenetic incongruities can be seen everywhere in the universal tree, from its root to the major branchings within and among the various taxa to the makeup of the primary groupings themselves"* [Woe98]. Though there are certainly minor changes in body structures, the sparsity of what are believed to be "transitional" forms, with none being unambiguous, is another problem if evolution is a small step at a time. Fossilized animals fall clearly within one of a limited number of basic body plans with clear morphological differences [Hal96]. Paleontologist and evolutionary biologist Henry Gee notes, *"To take a line of fossils and claim that they represent a lineage is not a scientific hypothesis that can be tested, but an assertion that carries the same validity as a bedtime story—amusing, perhaps even instructive, but not scientific"* [Gee99]. *"The animal body plans (as represented in the fossil record) do not grade imperceptibly one into another, either at a specific time in geological history or over the course of geological history. Instead, the body plans of the animals characterizing the separate phyla maintain their distinctive morphological and organizational features and thus their isolation from one another, over time"* [Mey03p333]. Dawkins writes concerning the invertebrate phyla fossils: *"It is as though they were just planted there, without any evolutionary history"* [Daw96Bp229].

"Evolution...must be gradual when it is being used to explain the coming into existence of complicated, apparently designed objects, like eyes... Without gradualness in these cases, we are back to miracle, which is simply a synonym for the total absence of explanation" [Daw95]. *"The problem of how eyes have developed has presented a major challenge to the Darwinian theory of evolution by natural selection. We can make many entirely useless experimental models when designing a new instrument, but this was impossible for Natural Selection, for each step must confer some advantage upon its owner, to be selected and transmitted through the generations. But what use is a half-made lens? What use is a lens giving an image, if there is no nervous system to interpret the information? How could a visual nervous system come about before there was an eye to give it information? In evolution there can be no master plan, no looking ahead to form structures which, though useless now, will come to have*

importance when other structures are sufficiently developed. And yet the human eye and brain have come about through slow painful trial and error" [Gre72]. There continues to be much speculation on the origin of the eye, ranging from multiple independent (as many as 60) times [Koz08] to a single evolved eye from which all eyes evolved [Geh05]. Even the simplest light sensitive spot involves a large number of specialized proteins and molecules in an extremely complicated integrated system. If any one of those proteins or molecules is missing in even the simplest eye system, there is no vision. A vision study of *"combination of spectral measurements and genetic sequences revealed which DNA mutations -- that is, amino acid substitutions -- were responsible for the shift in peak wavelength. The results were surprising: Identical substitutions didn't always produce the same shift... The results show how hard it is to identify productive mutations and to predict their effect. Showing conclusively how the survival of the fittest plays out on the molecular level would require reconstructing not only the protein, but also the whole animal and its long-lost habitat"* [Yok08].

The trilobite eye is a good case study since the trilobites suddenly appeared in the Cambrian era in less than 5 million years with no record of ancestry. The trilobite eye is made of optically transparent calcium carbonate (calcite, the same material of its shell) with a precisely aligned optical axis that eliminates double images and two lenses affixed together to eliminate spherical aberrations [McC98, Gal00]. The advanced optics, including bifocality, have produced observations like: *"Trilobites had solved a very elegant physical problem and apparently knew about Fermat's principle, Abbe's sine law, Snell's laws of refraction and the optics of birefringent crystals"* [Cla75]. In a description of *"The advantage of good eye design"* (with a note using design *"as a lead-in to the parallels between the optic designs of humans and the remarkably evolved morphology of trilobites"*) *"the rigid trilobite doublet lens had remarkable depth of field (that is, allowed for objects both near and far to remain in relatively good focus) and minimal spherical aberration"* [Gon07]. Physicist Riccardo Levi-Setti observes: *"In fact, this optical doublet is a device so typically associated with human invention that its discovery in trilobites comes as something of a shock. The realization that trilobites developed and used such devices half a billion years ago makes the shock even greater. And a final discovery – that the refracting interface between the two lens elements in a trilobite's eye was designed in accordance with optical constructions worked out by Descartes and Huygens in the mid-seventeenth century – borders on sheer science fiction"* [Lev93p57]. *"The trilobites of the Cambrian already had a* *highly advanced visual system. In fact, so far as we can tell from the fossil record thus far discovered, trilobite sight was far and away the most advanced in Kingdom Animalia at the base of the Cambrian... the lenses of the eyes of living trilobites were unique, being comprised of inorganic calcite... There is no other known occurrence of calcite eyes in the fossil record.* [FM-trib].

"There are three recognized kinds of trilobite eyes...with the great majority of trilobites bearing holochroal eyes... characterized by close packing of biconvex lenses beneath a single corneal layer that covers all of the lenses. These lenses are generally hexagonal in outline and range in number from one to more than 15,000 per eye" [Geo-web]. Some believe the abathochroal and schizochroal trilobite eye types evolved from holochroal [Tho05], but there is no fossil evidence to indicate that, and all three types, as well as eyeless trilobites are found in the same strata. *"Rarely, trilobites may have visual surfaces of normal size, but which lack lenses. This confirms that visual surface growth must have been regulated separately from lens emplacement, and is a feature that cannot be accounted for by the existing developmental model"* [Tho05]. Paleontologist Niles Eldredge observed: *"These lenses—technically termed aspherical, aplanatic lenses — optimize both light collecting and image formation better than any lens ever conceived. We can be justifiably amazed that these trilobites, very early in the history of life on Earth, hit upon the best possible lens design that optical physics has ever been able to formulate"* [Eld76]. *"The design of the trilobite's eye lens could well qualify for a patent disclosure"* [Lev93p58].

The trilobite lens is particularly intriguing since the only other animal to use inorganic focusing material is man. The lens may be classified as a prosthetic device since it was non-biological, which also means the lens itself was not subject to Darwinian evolution since it has no DNA. The manufacturing and controlling of the lenses were obviously biological processes, with an unknown number of DNA-produced proteins for collecting and processing the raw materials to manufacture the precision lenses and create the refracting interface between the two lenses. The lenses do not decompose as any other animal's lens would, so they are subject to rigorous scientific investigation and determination of optical properties based on the actual lenses, from which inferences can be made as to their use. For example, since they are multi-focal with spherical aberration correction, it can be inferred that trilobites had very good visual acuity. Finding a trilobite lens without knowing its source would undoubtedly lead to the conclusion that the lens had been designed (re-read the preceding paragraphs if this is doubted). Since no immediate precursors of trilobites have been found, Darwinists are without any evidence as to how an organism with an eye as complex as a trilobite could have arisen, especially in such a relatively short time.

Concerning the Cambrian explosion, *"Of course, scientists wedded to a purely materialistic explanation will instinctively deny the very possibility of top-down intelligent causation. Yet we regularly employ precisely this mode of explanation, especially when we encounter the kinds of patterns and features that we see in the fossil record. Indeed, we see in the fossil record several distinctive features or hallmarks of designed systems, including: (1) a quantum or discontinuous increase in specified complexity or information; (2) a top-down pattern of innovation in which large-scale morphological disparity arises before small-scale diversity; (3) the persistence of structural (or "morphological") disparities between separate organizational systems; and (4) the discrete or simultaneous emergence of functionally integrated material parts within novel organizational body plans. When we encounter objects that manifest any of these several features and we know how they arose, we invariably find that a purposeful agent or intelligent designer played a causal role in their origin. Thus, when we encounter all these same features in the fossil record, we may infer—based upon established cause-and--effect relationships and uniformitarian principles—that the same kind of cause operated in the history of life. In other words, intelligent design constitutes the best, most causally adequate, explanation of the*

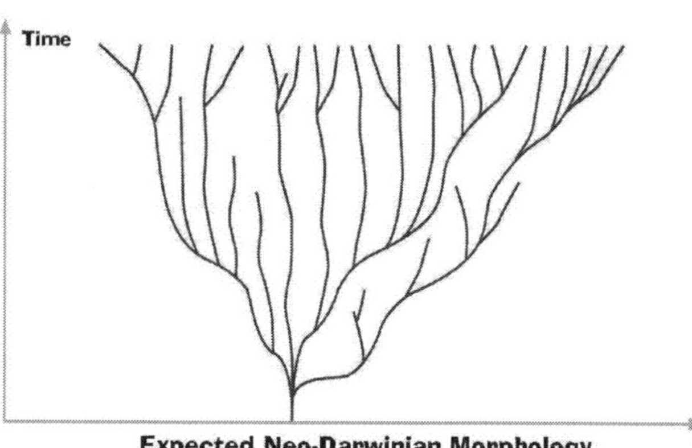

Expected Neo-Darwinian Morphology

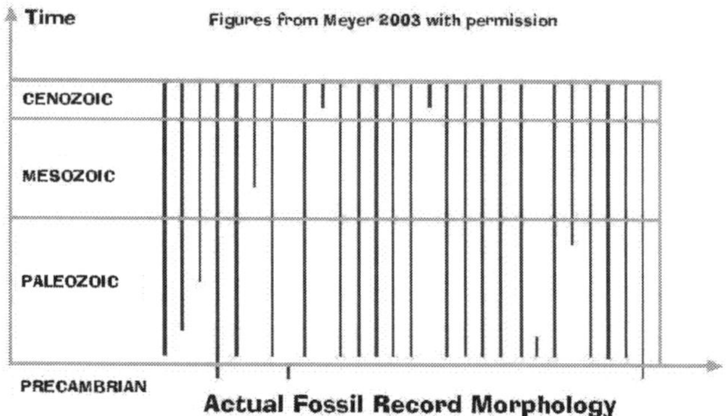

Actual Fossil Record Morphology

specific features of the Cambrian explosion, and the features of this explosion in turn attest to the activity and power of a purposeful intelligence" [Mey03p390].

Computer simulations and genetic algorithms [Bar57] are believed by many as evidence for the viability of neo-Darwinism. For example, Dawkins [Daw88] randomly "mutated" by random changes in 28 positions of letters and spaces:

```
"WDLTMNLT DTJBKWIRZREZLMQCO P"
```

to produce on the 43rd try:

```
"METHINKS IT IS LIKE A WEASEL"
```

Dawkins' program starts with a nonsense string and *"duplicates it repeatedly, but with a certain chance of random error – 'mutation' – in the copying. The computer examines the mutant nonsense phrases, the 'progeny' of the original phrase, and chooses the one which, however slightly, most resembles the target phrase"* [Daw88].

He knew the goal in advance and apparently kept mutations only if they became closer to that goal. Dawkins has not made his algorithm or source code available to the public (he is certainly invited to do so, as any scientist should, so that findings may be tested and verified), making what the program actually does difficult to ascertain. A partitioned search, mutating all unmatched positions simultaneously at each try, stopping further mutations at any position that becomes correct, would produce a quick solution.. With this scenario, the probability of success is high: : $1-(26/27)^{28} = 0.65$, for at least one desirable mutation on the first try, falling to no lower than 0.038 if only one position is incorrect (known mutation rates are under 10^{-8} [Smi89]). Dawkins probably used an unspecified mutation rate and an unspecified proximity search with an unspecified number of offspring to determine which progeny to retain. Many different search scenarios were evaluated [Ewe10] using the "best" guesses as to what Dawkins' algorithm might have been. Evolutionary algorithms out-performed a blind search, but were all relatively inefficient at extracting information. The result is that this exercise demonstrated nothing related to any understanding of evolution. It did show that an intelligently designed program with an intelligently designed target can work.

Ludwig [Lud93] sponsored an "Artificial Life" contest to find the shortest self-replicating program, with the winning program having 101 bytes. The probability of this program arising by chance is 256^{-101} or 10^{-243}. If 10^8 computers each make 10^7 trials/sec for 3×10^{22} trials/year, a solution becomes probable after 10^{220} years. If a suitable program were half as large, "only" 10^{99} years of processing would be necessary to make probable a self-replicating program by chance. These programs, like all computer programs, were intelligently designed and executed on designed platforms. Information, not random data, caused solutions. *"Based as it is on ideas, a computer is intrinsically an object of intelligent design. Every silicon chip holds as many as 700 layers of implanted chemicals in patterns defined with nanometer precision and then is integrated with scores of other chips by an elaborately patterned architecture of wires and switches all governed by layers of software programming written by human beings. Equally planned and programmed are all the computers running the models of evolution and 'artificial life' that are central to neo-Darwinian research. Everywhere on the apparatus and in the 'genetic algorithms' appear the scientist's fingerprints: the 'fitness functions' and 'target sequences.' These algorithms prove what they aim to refute: the need for intelligence and teleology* [targets] *in any creative process"* [Gil06].

Avida [Avida] is one of the most widely used [Ada02, Ofr03] artificial life programs, and is an auto-adaptive genetic system. The Avida system's concepts are similar to the Tierra program [Ray92], with a population of self-reproducing strings subjected to random (with Poisson distribution so the probability of k occurrences in interval λ is $\lambda^k e^{-\lambda}/k!$) mutations on a machine that can perform any calculation that any other programmable computer can perform (Turing- complete). The virtual computing organisms can together theoretically do any computation. The population of logic and math function (NOT, OR, AND, ADD, etc.) organisms *"adapts to the combination of an intrinsic fitness landscape (self-reproduction) and an externally imposed (extrinsic) fitness function provided by the researcher. By studying this system, one can examine evolutionary adaptation, general traits of living systems (such as selforganization), and other issues pertaining to theoretical or evolutionary biology and dynamic systems"* [Avida]. The virtual environment of Avida is initially seeded with a self-replicating human-designed program. This program and its descendants are subjected to random mutations which change instructions within their memory producing unfavorable, neutral, and favorable mutations. *"Mutations are qualified in a strictly Darwinian sense; any mutation which results in an increased ability to reproduce in the given environment is considered favorable. While it is clear that the vast majority of mutations will be unfavorable---typically causing the creature to fail to reproduce entirely---or else neutral, those few that are favorable will cause organisms to reproduce more effectively and thus thrive in the environment. Over time, organisms which are better suited to the environment are generated that are derived from the initial (ancestor) creature. All that remains is the specification of an environment such* *that tasks not otherwise intrinsically useful to self-reproduction are assimilated"* [Len03]. *"Avida makes it possible to watch the random mutation and natural selection of digital organisms unfold over millions of generations"* [Zim05]. The initial 50 instructions contains 15 instructions (not subject to mutations) for self-replication. Random mutations that increase genome length or generate useful math or logic functions (often using already evolved functions) on random inputs are rewarded with more CPU time (inheritable). *"While avida is clearly a genetic algorithm (GA) variation (to which nearly all evolutionary systems with a genetic coding can be reduced), the presence of a computationally (Turing) complete genetic basis differentiates it from traditional genetic algorithms. In addition, selection in avida more closely resembles natural selection than most GA mechanisms; this is a result of the implicit (and dynamic) co-evolutionary fitness landscape automatically created by the reproductive requirement. This co-evolutionary pressure classifies avida as an auto-adaptive system, as opposed to standard genetic algorithms (or adaptive) systems, in which the creatures have no interaction with each other. Finally, avida is an evolutionary system that is easy to study quantitatively yet maintains the hallmark complexity of living systems"* [AvidaMan]. Reread chapter 3 if you believe that such a computerized system really maintains the hallmark complexity of living systems.

Avida uses an unrealistically small genome, an unrealistically high mutation rate, unrealistic protection of the replication instructions, unrealistic energy rewards (external to genome) for genome size, and no capability for graceful functional degradation. It allows for arbitrary experimenter-specified selective advantages. *"Neglect of key factors or unrealistic parameter settings permit conclusions to be claimed which merely reflect what the decision maker intended a priori... the computer experiments reported using the Avida framework so far have not demonstrated that neo-Darwinian processes could have produced the necessary coding information to produce the hundreds of molecular machines found in natural cells"* [Tru04]. When more typical relative fitness values (1.01 - 1.1 range) were used, the logic functions found by chance mutations were removed from the genome, so that *"at the end of hundreds of thousands of generations (when the trials were terminated) no logic functions at all were present"* [Tru04]. Although Avida has been cited [Len03] as evidence against irreducible complexity, it is significant that when only the most complex function (EQU) was rewarded, it never appeared even in extremely long simulations [Len03]. EQU could only form from simpler functions, and when those are assigned no advantage, the complex function doesn't form. In short, Avida can be used to validate neo-Darwinism only if neo-Darwinism is taken as true (a programmed tautology).

Irreducible complexity is another problem that undirected Darwinian evolution has not addressed in any meaningful way. Behe defines irreducibly complex as *"a single system composed of several well-matched, interacting parts that contribute to basic function, wherein the removal of any one of the parts causes the system to effectively cease functioning. An irreducibly complex system cannot be produced directly... by slight, successive modifications of a precursor system, because any precursor to an irreducibly complex system that is missing a part is by definition non-functional"* [Beh96p39]. Behe's examples include the flagellum, aspects of blood clotting, closed circular DNA, telomeres (condensed DNA material), photosynthesis, and transcription regulation.

The bacterial flagellum is one of the best known examples of an irreducibly complex system [Beh96]. It is *"a microscopic rotary engine that contains parts known from human technology—such as a rotor, a stator, a propeller, a u-joint, and an engine—yet it functions at a level of efficiency that dwarfs any motor produced by humans. In any other context we would immediately recognize such an information-rich, integrated system as the product of intelligence"* [Lus08]. *"The bacterial flagellum of Salmonella typhimurium is the analogue of a man-made mechanical system. Its heart is a 15,000 revolutions per minute, reversible rotary motor powered by the proton-motive gradient across the cell's inner membrane. Each revolution consumes about 1000 protons. A drive shaft, held by a bushing in the outer membrane, transmits torque across the cell's envelope. Attached to the drive shaft, a universal joint enables the motor to drive the propeller, even when the drive shaft and propeller are not co-linear. A short junction joins the propeller to the drive shaft. The propeller, a long left-handed corkscrew, converts torque to thrust. A cap sits at the cell distal end of the filament. By electron microscopy, the motor associated*

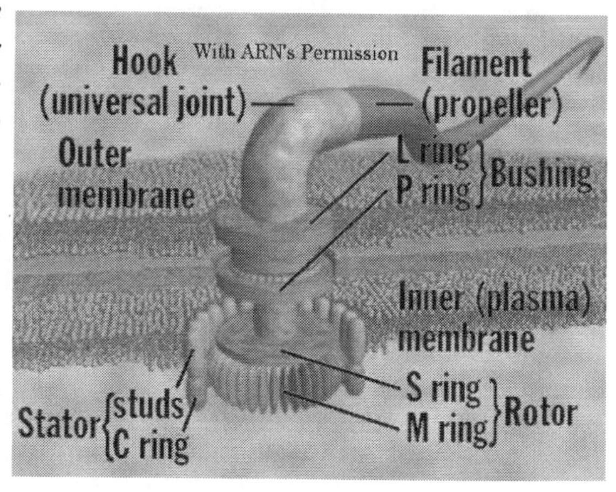

parts and the bushing are seen to be rings of subunits, whereas the drive shaft appears to be a helical assembly of subunits. About four dozen genes are needed to build the flagellum. Some are required for regulation of synthesis; some for export and assembly; some for the structure itself, and a few are of unknown function" [DeR95]. Over 40 distinct, carefully positioned proteins are used to construct the flagellum, including the various parts (including O-rings and bushings) and connectors to non-flagellum parts. *"The bacterial flagellar motor is an example of finished bio-nanotechnology, and understanding how it works and assembles is one of the first steps towards making man-made machines on the same tiny scale. The smallest man-made rotary motors so far are thousands of times bigger...* [a flagellum] *spins at up to 100,000 rpm and achieves near-perfect efficiency"* [Ber06B]. *"Flagellar proteins are synthesized within the cell body and transported through a long, narrow central channel in the flagellum to its distal (outer) end, where they self-assemble to construct complex nano-scale structures efficiently, with the help of the flagellar cap as the assembly promoter. The rotary motor, with a diameter of only 30 to 40 nm, drives the rotation of the flagellum at around 300 Hz, at a power level of 10^{-16} W with energy conversion efficiency close to 100 %. The structural designs and functional mechanisms to be revealed in the complex machinery of the bacterial flagellum could provide many novel technologies that would become a basis for future nanotechnology, from which we should be able to find many useful applications"* [Nam02].

If any one of 40 proteins is missing, a complete loss of motor function results [Dem01p253]. ATP is not used as an energy source, but instead, bacteria use energy from the flow of ions across membranes. *"One can only marvel at the intricacy, in a simple bacterium, of the total motor and sensory system which has been the subject of this review and remark that our concept of evolution by selective advantage must surely be an oversimplification. What advantage could derive, for example, from a 'preflagellum' (meaning a subset of its components), and yet what is the probability of 'simultaneous' development of the organelle at a level where it becomes advantageous?"* [Mac78] It has been noted that several proteins that are needed by the flagellum are also found in Type III Secretion Systems (TTSS) of bacteria, with the speculation that TTSS for poison injection is a precursor of the flagellum. Although 10 genes are shared, 40 (30 in the motor) genes are unique to the flagellum. Phylogenetic analysis of gene sequences *"suggest that flagellar motor proteins arose first and those of the pump came later... if anything, the pump evolved from the motor, not the motor from the pump"* [Min04]. The pump is used during flagellum construction since self-assembly is performed at the tip (distal end). Since each protein requires a prescriptive algorithm implemented in the DNA, the simultaneous "arising" of at least 30 such computer programs is difficult to justify from information science. When we consider 10^{-164} as the probability of even a simple life-compatible protein [Mey09p212] (even forming one is operationally falsified [Abe09U]), the probability of forming all 30 becomes 10^{-4920}.

Note also that new proteins typically *"require multiple new protein folds, which in turn require long stretches of new protein sequence.... the vast set of possible proteins that could conceivably be constructed by genetic mutations is far too large to have actually been sampled to any significant extent in the history of life. Yet how could the highly incomplete sampling that has occurred have been so successful? How could it have located the impressive array of protein functions required for life in all its forms, or the comparably impressive array of protein structures that perform those functions?... If we take 300 residues as a typical chain length for functional proteins, then the corresponding set of amino acid sequence possibilities is unimaginably large, having 20^{300} ($= 10^{390}$) members...* [Mutational paths to create] *new folds will inevitably destabilize the original fold before producing the new one... the sampling problem— the impossibility of any evolutionary process sampling anything but a miniscule fraction of the possible protein sequences... greatly strengthened the case that the sampling problem is real and that it does present a serious challenge"* [Axe10].

Dembski [Dem04] points out the daunting probabilistic hurdles Darwinians face when attempting coordination of needed successive evolutionary changes needed for biomechanical machines that are irreducibly complex including availability and synchronization of parts, elimination of interfering cross-reactions and proper interfacing with other components, and ordering the assembly so that a functioning system results. Bracht [Bra03], Behe [Beh00], Williams [WilWeb], and Dembski [Dem04] have adequately addressed issues raised by skeptics concerning irreducible complexity, especially that of the flagellum. It should be noted that Yockey's statement *"the protein sequences that compose living organisms are not random or 'irreducibly complex'"* [Yoc05p179], is technically true considering only Shannon information and not functional information. Irreducible complexity, as used by Behe, deals with forming new structures (an engineering problem, rather than mathematical puzzle), along with the associated proteins that require considerable functional information.

Irreducible complexity may also be found in the DNA/RNA/enzyme system.

"Nucleic acids are synthesized only with the help of proteins, and proteins are synthesized only if their corresponding nucleotide sequence is present. It is extremely improbable that proteins and nucleic acids, both of which are structurally complex, arose spontaneously in the same place at the same time. Yet it also seems impossible to have one without the other. And so, at first glance, one might have to conclude that life could never, in fact, have originated by chemical means" [Org94]. Sexual reproduction may also be an example of irreducible complexity since evolutionary theory has failed to explain its origin. The ubiquitous nature of sexual reproduction (whether in plants or animals) is based on gender differences and defies any naturalistic explanation. How could a female member of a species evolve to produces eggs and be internally equipped to nourish a growing embryo, while at the same time a male member evolve to produce viable sperm, with each gamete (sperm or egg) evolving to contain half the number of chromosomes? Sex's *"disadvantages seem to outweigh its benefits. After all, a parent that reproduces sexually gives only one-half its genes to its offspring, whereas an organism that reproduces by dividing passes on all its genes. Sex also takes much longer and requires more energy than simple division. Why did a process so blatantly unprofitable to its earliest practitioners become so widespread?"* [Sch84] *"Despite decades of speculation, we do not know. The difficulty is that sexual reproduction creates complexity of the genome and the need for a separate mechanism for producing gametes. The metabolic cost of maintaining this system is huge, as is that of providing the organs specialized for sexual reproduction"* [Mad98]. *"Sex is a puzzle that has not yet been solved; no one knows why it exists"* [Rid01].

Recombinant DNA [Coh73], the "creation" of new life forms in the laboratory, may raise questions whether such creations support evolution. The artificial DNA is engineered through combination or insertion of one or more existing DNA strands to produce DNA sequences not normally occurring together. The modified or new traits are designed for a specific purpose, such as immunity or a new fuel [Ven08]. No new net information is generated, as existing information is used to form the modified information, with some of the original information lost in the new DNA. As an analogy, this book has to this point not created any new information structures (words). If "juxtaword" is used in this paragraph, the reader may recognize the meaning. It's unclear how many words can be juxtaworded (using the principle than any noun can be verbbed), or the limits of juxtawording, or if multiple justawords may be juxtaworded. If people like the use of "juxtaword," it could spread to the population in general, becoming a permanent trait. This is an example of a minor modification that could have been produced by juxtaposing the first five letters of the information unit "juxtapose" with the information unit "word." Note that the "pose" information is missing in "juxtaword." Information units are analogs to genes, with the letters being analogs to the genetic alphabet. If a totally new information unit "swervozt" is randomly generated from the alphabet, it's extremely unlikely that it would be replicated throughout the population, since it was produced without purpose, and in fact contains no functional information (although the Shannon information is higher than the normal word due to the unpredictability of the letters).

The "artificial genome" produced by the Venter Institute was unique in genetic engineering in that a genome that matched (except for a few additions, such as identifying watermarks) an existing bacterium was engineered from existing parts (all parts originated from life). This genome replaced the genome in another bacterium, leaving all the computing machinery, operating systems, and other components of that cell intact. To put this in perspective, copying the executable binary code of a program to a USB drive and then inserting the drive into another computer is not creating a system "from scratch." The new genome was engineered using computers *"starting from digitized genome sequence information"*[Gib10], verifying the digital computing nature of life. The complexity and specificity of life's information is highlighted by *"obtaining an error-free genome that could be transplanted into a recipient cell to create a new cell controlled only by the synthetic genome was complicated and required many quality control steps. Our success was thwarted for many weeks by a single base pair deletion in the essential gene dnaA. One wrong base out of over one million in an essential gene rendered the genome inactive"* [Gib10]. One of the things this research supports is the idea that (at least for the two closely-related bacteria involved) life uses common operating systems, programming languages, and devices (otherwise the programs for one machine wouldn't execute on another).

While there is evidence that life increased in complexity with many new body structures over time, there is no scientific evidence that the increase in information required to make those changes could have occurred by undirected natural processes, as is usually assumed. The microevolutionary changes that have been observed seem to have a very limited range to affect form or function, whereas the changes required for macroevolutionary changes have not been observed or are detrimental. *"Loci that are obviously variable within natural populations do not seem to lie at the basis of many major adaptive changes, while those loci that seemingly do constitute the foundation of many if not most major adaptive changes are not variable"* [McD83]. Many use the genetic or morphological similarities as support of common ancestry. Others see such features as evidence of common design since the same proteins (and hence the same DNA sequences) are used by many species, and many physical needs are similar. For example, in designing a new device or program, a good engineer reuses previously successful implementations wherever appropriate.

Evolutionists that support ID see the fitness functions of life and the target species intelligently placed within life (as is done in artificial life pro-grams), so that evolution becomes a subset of ID. *"There is no inherent conflict between evolutionary theory and the theory of intelligent design. However, ...that evolution is driven by natural selection acting on random mutations, an unpredictable and purposeless process that 'has no discernable direction or goal, including survival of a species'... that intelligent design directly challenges"* [Wel07]. *"ID doesn't hold that the neo-Darwinist theory of evolution is wholly and straightforwardly wrong; but it does suggest that it is only a partially successful theory that needs to be subsumed by a wider explanatory hypothesis of design, somewhat as Newton's Laws were subsumed by Einstein's"* [WilWeb].

The neo-Darwinian model for evolution is increasingly being challenged by scientists. The observed *"multi-character changes are fundamentally different from the slowly accumulating small random variations postulated in Darwinian and neo-Darwinian theory... One of the traditional objections to Darwinian gradualism has been that it is too slow and indeterminate a process to account for natural adaptations, even allowing for long periods of random mutation and selection"* [Sha10]. *"Evolutionary-genomic studies show that natural selection is only one of the forces that shape genome evolution and is not quantitatively dominant, whereas non-adaptive processes...undermine the Tree of Life concept. An adequate depiction of evolution requires the more complex concept of a network or 'forest' of life... Evolutionary genomics effectively demolished the straightforward concept of the TOL by revealing the dynamic, reticulated character of evolution"* [Koo09]. *"Because biological systems are complex, a unified theoretical framework that coordinates, integrates, and even partially embeds a plurality of theories about systems is required to capture and manage this complexity... any single theoretical abstraction would lead to dangerous reifications... Ignoring key and legitimate abstractions can lead to limited understand-*

ing, short-sightedness, and the stalling of theoretical and empirical research" [Win08]. *"Natural selection is the long-term result of molecular copying and would be the sole mechanism of evolution if copying were the sole basic mechanism of life. But there are two distinct molecular mechanisms at the basis of life, copying ... leads in the long run to natural selection and coding to natural conventions* [biosemiotics], *which means that evolution took place by two distinct mechanisms. Natural selection produces new objects by modifying previous ones, whereas natural conventions bring absolute novelties into existence"* [Bar08B]. New structures and new species would be the result of new arbitrary natural coding systems.

It is extremely important to realize that neo-Darwinism is not a mechanism for evolutionary change! A mechanism would explain how new structures develop. Each structure is the result of multiple proteins, each requiring an instantiated prescriptive algorithm for its production. The finished structures of an organism that can contribute to survival selection are determined by the genetic changes. Darwinism offers no feasible mechanism for specifying the algorithms or other functional information at the genomic level. Darwinism asserts that the fittest will survive, and that those that survive are the fittest, which is tautology that can be stated "those that survive, survive." You are now able to evaluate the claim that anyone doubting Darwinism *"is ignorant, stupid, or insane"* [Daw89]. Darwinism is insufficient to explain the observations.

"Creation of new information is habitually associated with conscious activity" [Qua64]. We know from artificial life programs that no positive evolutionary progress is made without appropriate intelligently designed fitness functions and targets. We also know that functional complexity of organisms has increased from the lowest strata to the present, with many unique structures that would require increased information in the genomes. The formation probability ratio of the simplest life (267,000 bits of functional information) to the human genome (>10^8 bits) would be $2^{10000000} = 10^{3000000}$. But somehow, over 10^7 bits of functional information must have been injected into the genome. It is relatively easy to see how a genome can increase in length via insertion or replication, but such mutations would add almost no information; e. g, the last three repeats in "Anythingnew?Anythingnew?Anythingnew?Anythingnew?," add no Shannon or functional information (except for a repeat count of 4). There is no known naturalistic mechanism for producing any increase in information, let alone information of the magnitude necessary to evolve from the root to the top of life's tree. Science needs to seek verifiable answers, not dogmatic pronouncements.

7 Going Where Data Lead

"The scientific attitude has usually been characterised as a commitment to following the evidence wherever it leads. That does not look like promising ammunition for someone pushing an official policy of refusing to allow science to follow evidence to... design no matter what the evidence turns out to be... it commits science to either having to deliberately ignore major (possibly even observable) features of the material realm or having to refrain from even considering the obvious and only workable explanation, should it turn out that those features clearly resulted from [intelligent] *activity... any imposed policy of naturalism in science has the potential not only of eroding any self-correcting capability of science but of preventing science from reaching certain truths. Any imposed policy of methodological naturalism will have precisely the same potential consequences"* [Rat00]. *"By making our explanation into the definition of the condition to be explained, we express not scientific hypothesis but belief. We are so convinced that our explanation is true that we no longer see any need to distinguish it from the situation we were trying to explain. Dogmatic endeavors of this kind must eventually leave the realm of science"* [Bra85]. *"I believed – took it as a given – that my science textbooks represented the best scientific knowledge available at that time. It was only when I was finishing my Ph.D. in cell and development biology, however, that I noticed what at first I took to be a strange anomaly. The textbook I was using prominently featured drawings of vertebrate embryos – fish, chickens, humans, etc. – where similarities were presented as evidence for descent from a common ancestor. Indeed, the drawings did appear very similar. But I'd been studying embryos for some time, looking at them under a microscope. And I knew that the drawings were just plain wrong. I re-checked all my other textbooks. They all had similar drawings, and they were all obviously wrong"* [Wel01].

"Science" has its root in the Latin word "Scientia," meaning knowledge or truth. *"Science... accumulated and accepted knowledge that has been systematized and formulated with reference to the discovery of general truths or the operation of general laws: knowledge classified and made available in work, life, or the search for truth: comprehensive, profound, or philosophical knowledge; especially knowledge obtained and tested through the use of the scientific method"* [Web93]. Nobel prize winner Linus Pauling said: *"Science is the search for truth, the effort to understand the world; it involves the rejection of bias, of dogma, of revelation, but not the rejection of morality...*

One way in which scientists work is by observing the world, making note of phenomena, and analyzing them" [PauWeb]. It would be nice to believe that presuppositions and biases are not involved with science, and that scientists would follow the evidence, wherever it took them. It is healthy and proper to thoroughly examine claims that purport to be scientific because anyone can speculate anything, but that doesn't make it true. What should not be tolerated is refusal to evaluate the scientific merits because of philosophical or theological views. For example, biologist Richard Lewontin writes *"Our willingness to accept scientific claims that are against common sense is the key to an understanding of the real struggle between science and the supernatural. We take the side of science in spite of the patent absurdity of some of its constructs, in spite of its failure to fulfill many of its extravagant promises of health and life, in spite of the tolerance of the scientific community for unsubstantiated just-so stories, because we have a prior commitment, a commitment to materialism. It is not that the methods and institutions of science somehow compel us to accept a material explanation of the phenomenal world, but, on the contrary, that we are forced by our a priori adherence to material causes to create an apparatus of investigation and a set of concepts that produce material explanations, no matter how counter-intuitive, no matter how mystifying to the uninitiated. Moreover, that materialism is absolute, for we cannot allow a Divine Foot in the door"* [Lew97]. Dawkins writes: *"Designoid objects look designed, so much so that some people – probably, alas, most people – think that they are designed. These people are wrong... the true explanation – Darwinian natural selection – is very different"* [Daw96Cp4-5]. Dawkins illustrates that "designoid" objects superficially appear a bit like they are designed, and differ from those that are clearly designed. He compares a Kennedy-like resemblance on a hill to Mt. Rushmore. He goes on to assert that no biological organisms are designed, but are merely "designoids" (despite life's cybernetic complexity).

Scientific dogmatism is counterproductive, even if widespread. Jonathan Wells [Wel00] writes of 10 evolutionary "Icons" that even evolutionists admit are misrepresentations, yet continue to be used as evidence of evolution. These icons include the Miller-Urey experiment, Darwin's tree of life, limb homology, Haeckel's embryos, archaeopteryx, peppered moths, Darwin's finches, 4-winged fruit flies, horse evolution, and ape-to-man evolution. *"Hearing how key Design advocates came to their current view, it became clear that their entry into the movement stemmed from intellectual or scientific – not religious – reasons... Several of the founders frequently relate a vivid tale of how they previously had assumed the validity of Darwinian scenarios and were later shocked to discover major weaknesses in the case for Darwinism. Typically this intellectual epiphany leads to further reading and research, which cements the new radical doubt about the theory's plausibility"* [Woo03].

Unfortunately, many scientists and others equate ID with "Creationism," and refuse to even consider scientific evidence that supports ID. This author has experienced such bias first-hand in an ID debate at a public college, wherein the opponent continually made accusations of "bringing God into the debate," whereas, in actuality, the only mention of "God" was in Fred Hoyle's (an atheist at that time) explanation of the enzymes for life probability calculation: *"the intelligence which assembled the enzymes did not itself contain them...which by no means need be God, however"* [Hoy81Ep139]. By contrast, when an ID-friendly poster [Joh04] was presented at a Bioinformatics conference, the reception was friendlier than anticipated, with dozens of scientists expressing support or non-negative interest (less than a third were negative) in the concepts. It seems that scientists who have been actually investigating the vast information of life are more willing to consider models that many mainstream scientists find objectionable. This book is about science, and it is hoped that it will be evaluated based on science, and not philosophical beliefs. This chapter is meant to point out beliefs that prevent many scientists from engaging in an honest evaluation of the scientific content. The purpose is not to advocate any particular belief. It is unfortunate that such a chapter is necessary.

The best known ID proponent is the Discovery Institute, so it is useful to examine ID based on what they say it is (as opposed to what someone not connected to ID says ID is). *"Intelligent design refers to a scientific research program as well as a community of scientists, philosophers and other scholars who seek evidence of design in nature. The theory of intelligent design holds that certain features of the universe and of living things are best explained by an intelligent cause, not an undirected process such as natural selection. Through the study and analysis of a system's components, a design theorist is able to determine whether various natural structures are the product of chance, natural law, intelligent design, or some combination thereof. Such research begins by observing the types of information produced when intelligent agents act. Scientists then seek to find objects which have those same types of informational properties which we commonly know come from intelligence. Intelligent design has applied these scientific methods to detect design in irreducibly complex* *biological structures, the complex and specified information content in DNA, the life-sustaining physical architecture of the universe, and the geologically rapid origin of biological diversity in the fossil record during the Cambrian explosion approximately 530 million years ago... The theory of intelligent design is simply an effort to empirically detect whether the 'apparent design' in nature acknowledged by virtually all biologists is genuine design (the product of an intelligent cause) or is simply the product of an undirected process such as natural selection acting on random variations"* [IdWeb].

Although many aspects of ID are compatible with various theological views, ID makes no claim concerning the identity of the agent(s) causing the empirically detectable marks of intelligence. There are undoubtedly philosophical and theological ramifications, but since those fall outside the realm of empirical science, ID proponents make no pronouncements in support of any particular view. The US Supreme Court has held that the implications of material alone do not make a religion even though those implications *"coincide or harmonize with the tenets of some or all religions"* [Sup80]. It also ruled *"The Establishment Clause stands at least for the proposition that when government activities touch on the religious sphere, they must be secular in purpose, evenhanded in operation, and neutral in primary impact"* [Sup71]. A recent US Appeals Court ruling rejected the claim that *"Texas Education Agency's ('TEA') neutrality policy constitutes an establishment of religion, in violation of the First Amendment's Establishment Clause. Because we find no evidence to support the conclusion that the principal or primary effect of TEA's policy is one that either advances or inhibits religion, we conclude that the policy does not violate the Establishment Clause. As such, we affirm the decision of the district court"* [App10]. The National Center for Science Education supported the suit, claiming that the TEA policy was endorsing "Creationism" (which it wasn't, but any evidence potentially bringing Darwinism into question is thought by many naturalists to be a religious stance). Excellent legal reviews of court rulings are available [Lus09, Cal09].

Some may object to ID because of religious compatibility, which should mean they should be equally adamant against Darwinism, since that is compatible with Atheism, Secular Humanism, and Naturalism, which are definitely religions, as affirmed by the Supreme Court: *"Among religions in this country which do not teach what would generally be considered a belief in the existence of God are Buddhism, Taoism, Ethical Culture, Secular Humanism, and others"* [Sup61]. It also indicated that *"religious beliefs... are based... upon a faith, to which all else is subordinate or upon which all else is ultimately dependent.... Some believe in a purely personal God, ... others think of religion as a way of life"* [Sup65]. This Supreme Court criterion makes "physicality is all there is" a religious belief since it cannot be proven and all else is subordinate to it. A US Appellate court also affirmed that *"Atheism is religion, and the group... was religious in nature even though it expressly rejects a belief in a supreme being"* [App05]. Because of these court rulings, one must use care when presenting any speculative purely physical naturalistic scenario, since transgression could be an establishment violation (with the same warnings as teaching Creationism would bring). Presenting only verifiable science, avoiding unverified speculations, should be safe.

Sometimes science seems openly hostile to anything perceived as religious. Francisco Ayala, a former president of the American Association for the Advancement of Science, has stated: *"The functional design of organisms and their features would therefore seem to argue for the existence of a designer. It was Darwin's greatest accomplishment to show that the directive organization of living beings can be explained as the result of a natural process, natural selection, without any need to resort to a Creator or other external agent"* [Aya94p323]. Atheist and science historian Will Provine writes: *"As the creationists claim, belief in modern evolution makes atheists of people. One can have a religious view that is compatible with evolution only if the religious view is indistinguishable from atheism"* [Pro99]. Anything other than naturalism is dismissed by biologist George Wald by admitting *"I will not believe that philosophically because I do not want to believe in God. Therefore, I choose to believe in that which I know is scientifically impossible: spontaneous generation arising to evolution"* [Wal58]. Michael Ruse states that evolution is *"a full-fledged alternative to Christianity... Evolution is a religion. This was true of evolution in the beginning, and it is true of evolution still today"* [Rus00]. On rare occasions, the divine even seems acceptable, such as that in *Science: "The fact that the universe exhibits many features that foster organic life -- such as precisely those physical constants that result in planets and long-lived stars -- also has led some scientists to speculate that some divine influence may be present"* [Eas97].

Often Darwinism becomes a de facto religion. "*Directed by all-powerful selection, chance becomes a sort of providence, which, under the cover of atheism, is not named but which is secretly worshiped... To insist, even with Olympian assurance, that life appeared quite by chance and evolved in this fashion, is an unfounded supposition which I believe to be wrong and not in accordance with the facts"* [Gra77p107]. *"Our theory of evolution has become... one which cannot be refuted by any possible observations. Every conceivable observation can be fitted into it. It is thus 'outside of empirical science' but not necessarily false. No one can think of ways in which to test it. Ideas... have become part of an evolutionary dogma accepted by most of us as part of our training"* [Ehr67]. Wolfgang Smith (mathematician, physicist, and philosopher of science) writes: *"I am convinced, moreover, that Darwinism, in whatever form, is not in fact a scientific theory, but a pseudo-metaphysical hypothesis decked out in scientific garb. In reality the theory derives its support not from empirical data or logical deductions of a scientific kind but from the circumstance that it happens to be the only doctrine of biological origins that can be conceived with the constricted world view to which a majority of scientists no doubt subscribe"* [Smi92]. Atheist and evolutionary biologist David Sloan Wilson admits, *"many scientific theories of the past become weirdly implausible...* [and] *are a greater cause for concern because they do a better job of masquerading as factual reality. Call them stealth religions"* [Wil07]. After a lecture, atheist biologist Richard Dawkins is reported to have *"admitted in a Q&A that followed of being 'guilty' of viewing Darwinism as a kind of religion"* [Maz10p97].

A supernatural designer is erroneously believed by many to be required by ID. It is worthwhile to look at some pertinent statements, such as by evolutionist Michael Denton. *"It is important to emphasize at the outset that the argument presented here is entirely consistent with the basic naturalistic assumption of modern science – that the cosmos is a seamless unity which can be comprehended ultimately in its entirety by human reason and in which all phenomena, including life and evolution and the origin of man, are ultimately explicable in terms of natural processes"* [Den98pxviii]. And later in that book: *"Evolution was accepted in the nineteenth century not because it explained everything perfectly but because it accounted for the facts better than any other theory...The idea that the cosmos is a unique whole with life and mankind as its end and purpose makes sense and illuminates all our current scientific knowledge. It makes sense of the intricate synthesis of carbon in the stars, of the constants of physics, of the properties of water, of the cosmic abundance of the elements... No other world view comes close. No other explanation makes as much sense of all the facts"* [Den98p385]. Agnostic Bill Schultz writes *"'intelligent design' arguments do not in any way support any assertion of the existence of some supernatural deity... it is entirely* *consistent with what we currently know about our universe for some external but natural intelligence to have 'designed' our universe to be what it is. Such intelligence would be mighty indeed, but it would still be just another powerful alien, or a group of such aliens, and not in any way a god or gods"* [Sch99].

Fred Hoyle was an atheist when he wrote *"The enormous information content of even the simplest living systems... cannot in our view be generated by what are often called 'natural' processes... There is no way in which we can expect to avoid the need for information, no way in which we can simply get by with a bigger and better organic soup, as we ourselves hoped might be possible... The correct position we think is... an intelligence, which designed the biochemicals and gave rise to the origin of carbonaceous life... This is tantamount to arguing that carbonaceous life was invented by noncarbonaceous intelligence"* [Hoy81E]. In a statement prepared for court, Wickramasinghe explains *"even if chemical barriers for the linkages are artificially and miraculously removed, the really vast improbability of 1 in 10^40,000 poses a serious dilemma for the whole of evolutionary science. Life could not be an accident, not just on the Earth alone, but anywhere, anywhere at all in the Universe. The facts as we now see them point to one of two distinct conclusions: an act of deliberate creation, or an indelible permanence of the patterns of life in a Universe that is eternal and boundless. For those who accept modern cosmological views as gospel truth, the latter alternative might be thought unlikely, and so one might be driven inescapably to accept life as being an act of deliberate creation. Creation would then be brought into the realm of empirical science. The notion of a creator placed outside the Universe poses logical difficulties, and is not one to which I can easily subscribe. My own philosophical preference is for an essentially eternal, boundless Universe, wherein a creator of life somehow emerges in a natural way. My colleague, Sir Fred Hoyle, has also expressed a similar preference. In the present state of our knowledge about life and about the Universe, an emphatic denial of some form of creation as an explanation for the origin of life implies a blindness to fact and an arrogance that cannot be condoned"* [Wic81].

Some believe that nature itself possesses intelligence capable of design. For example, Shapiro's natural genetic engineering *"employs a combinatorial search process based upon DNA modules that already possess functionality... Such a cognitive component is absent from conventional evolutionary theory because 19th and 20th century evolutionists were not sufficiently knowledgeable about cellular response and control networks... It also answers the objections to conventional theory raised by intelligent design advocates, because evolution by natural genetic engineering has the capacity to generate complex novelties"* [Sha10]. Even though he denies ID, what he proposes is precisely ID! A cell would have a cognitive mind capable of doing a combinatorial search of possible outcomes to design the best mutation before mutating.

It is clear from explicit statements that ID is not a religion, and makes no claim as to the identity of the design agent(s). Obviously, ID proponents have the freedom of religion allowed by the country of residence, and many are theists, but those beliefs should not detract from the factual scientific evidence. It should also be noted that the specific "purpose" of design in nature is not scientifically knowable (even though design implies purpose), and therefore believing a specific purpose is not required for accepting the detectable evidences of design. Many, if not most, ID proponents do hold "purpose" as philosophical or theological (not scientific) beliefs compatible with ID evidence. Many have been motivated by the ID evidence to seek answers to important philosophical and theological questions that cannot be answered by empirical science. Those philosophical and theological beliefs influence many areas of personal life as well, such as philanthropy, responsibility, and civility.

Undirected natural processes as an absolute belief would also have impact on society. *"All appearances to the contrary, the only watchmaker in nature is the blind forces of physics, albeit deployed in a very special way. A true watchmaker has foresight: he designs his cogs and springs and plans their interconnections, with a future purpose in his mind's eye. Natural selection, the blind, unconscious, automatic process which Darwin discovered, and which we now know is the explanation for the existence and apparently purposeful form of all life, has no purpose at all"* [Daw96Bp5]. *"The time has come to take seriously the fact that we humans are modified monkeys, not the favored Creation of a Benevolent God on the Sixth Day. In particular, we must recognize our biological past in trying to understand our interactions with others. We must think again especially about our so-called 'ethical principles.' The question is not whether biology—specifically, our evolution—is connected with ethics, but how. As evolutionists, we see that no justification of the traditional kind is possible. Morality, or more strictly our belief in morality, is merely an adaptation put in place to further our reproductive ends. Hence the basis of ethics does not lie in God's will... ethics as we understand it is an illusion fobbed off on us by our genes to get us to cooperate. It is without external grounding... Ethics is illusory inasmuch as it persuades us that it has an objective reference. This is the crux of the* *biological position. Once it is grasped, everything falls into place"* [Rus91].

Since Darwinism has no purpose and survival of the fittest is "law," humans should not be concerned with endangered species, since any such species isn't fit to survive. Besides, new species will evolve that will be fit. Since humans are "just" animals, there is no such thing as "human rights," and hence no such thing as a violation of those rights. Democracy is both inefficient and unnatural, so dictatorships should be the prevailing world order, with the strongest at the top. Anything that enhances one's ability to survive should be allowed, according to Darwinism, so that laws regulating behaviors are inappropriate, including those prohibiting murder, robbery, and rape (after all, an alpha male should be able to pass on his genes with as many females as possible). Defective humans should be killed before being allowed to reproduce so the species remains genetically pure. Hitler used his belief in Darwinism to justify extermination of those he considered below the superior Arian race [Kei47, Wei04, Exp08], in line with the full title of Darwin's book "On the Origin of Species by Means of Natural Selection, or the Preservation of Favoured Races in the Struggle for Life."

Thankfully, most people are repulsed by those views, which means few people really believe that undirected natural processes are the source of humanity. Even Richard Dawkins admits *"there are two reasons why we need to take Darwinian natural selection seriously. Firstly, it is the most important element in the explanation for our own existence and that of all life. Secondly, natural selection is a good object lesson in how NOT to organize a society. As I have often said before, as a scientist I am a passionate Darwinian. But as a citizen and a human being, I want to construct a society which is about as un-Darwinian as we can make it. I approve of looking after the poor (very un-Darwinian). I approve of universal medical care (very un-Darwinian)"* [Daw08L]. This is in contrast to his statement *"I want to persuade the reader, not just that the Darwinian world-view happens to be true, but that it is the only known theory that could, in principle, solve the mystery of our existence"* [Daw96Bpxiv]. It is intriguing that one of the strongest, if not the strongest, voices supporting Darwinism really doesn't follow those beliefs to where they lead. When attempting to look up [Web87] "integrity" some years ago, it wasn't found until the meaning of "integer" was examined, where "integrity" was found embedded. This was interesting to a computer science teacher who teaches that an integer is an unfractionated whole.

People, including scientists, are free to hold whatever beliefs they desire, but when actions taken as a result of those beliefs cause harm to others, corrective action is needed. As a classic example of science run amuck, consider the case of Richard Sternberg, who approved an ID-friendly paper by Stephen Meyer, whose conclusion included *"What natural selection lacks, intelligent selection--purposive or goal-directed design--provides. Rational agents can arrange both matter and symbols with distant goals in mind. In using language, the human mind routinely 'finds' or generates highly improbable linguistic sequences to convey an intended or preconceived idea...The causal powers that natural selection lacks--almost by definition--are associated with the attributes of consciousness and rationality--with purposive intelligence. Thus, by invoking design to explain the origin of new biological information, contemporary design theorists are not positing an arbitrary explanatory element unmotivated by a consideration of the evidence. Instead, they are positing an entity possessing precisely the attributes and causal powers that the phenomenon in question requires as a condition of its production and explanation"* [Mey04]. The publication of this ID-friendly paper as science created an uproar in the scientific community.

Sternberg's short fact statement on this incident includes *"In the case of the Meyer paper, I followed all the standard procedures for publication in the Proceedings of the Biological Society of Washington... It was my prerogative to choose the editor who would work directly on the paper, and as I was best qualified among the editors I chose myself, something I had done before in other appropriate cases. In order to avoid making a unilateral decision on a potentially controversial paper, however, I discussed the paper on at least three occasions with another member of the Council of the Biological Society of Washington (BSW), a scientist at the National Museum of Natural History. Each time, this colleague encouraged me to publish the paper despite possible controversy. The Meyer paper underwent a standard peer review process by three qualified scientists, all of whom are evolutionary and molecular biologists teaching at well-known institutions. The reviewers provided substantial criticism and feedback to Dr. Meyer, who then made significant changes to the paper in response. Subsequently, after the controversy arose, Dr. Roy McDiarmid, President of the Council of the BSW, reviewed the peer-review file and concluded that all was in order"* [Ste08].

Even though there are several ID-friendly peer-reviewed publications, it is interesting that many scientists use the sparsity of peer-reviewed publications that are ID-friendly as "evidence" that ID isn't really science. Then, when a paper that went through an extraordinary review process actually does get published, the editor gets harassed for allowing the paper, despite the fact that more than just the normal peer reviewer process approved publication. University Professors have been fired and researchers have lost funding as a result of writing or speaking positively about ID. [Exp08, Ber08, Cro10] It is understandable that scientists would avoid research that would cause their funding to dry up or for them to lose their jobs. It is also understandable that they will do almost anything to avoid having to admit they, in all their intelligence, have been duped into believing unsubstantiated speculations (as this author had been). But scientific integrity and believability is at stake. Scientists need to look at the facts to see what conclusions can be drawn.

The scientific "establishment" has also inhibited other (besides ID) scientific views. Dissenters to big bang cosmology note that opposing view *"development has been severely hampered by a complete lack of funding. Indeed, such questions and alternatives cannot even now be freely discussed and examined... all the peer-review committees that control them are dominated by supporters of the big bang. As a result, the dominance of the big bang within the field has become self-sustaining, irrespective of the scientific validity of the theory"* [Ope04]. *"Science is a process within a community of people looking for truth; string theory is just a faith-based science... for twenty years it has soaked up most of the funding, attracting some of the best scientific minds, and penalizing young physicists for pursuing other avenues"* [Smo07].

To expound some scientifically mainstream views, Pagels writes that so powerful is the scientific-experimental *"method that virtually everything scientists know about the natural world comes from it. What they find is that the architecture of the universe is indeed built according to invisible universal rules, what I call the cosmic code-the-building code of the Demiurge. Examples of this universal building code are the quantum and relativity theory, the laws of chemical combination and molecular structure, the rules that govern protein synthesis and how organisms are made, to name but a few. Scientists in discovering this code are deciphering the Demiurge's hidden message, the tricks he used in creating the universe. No human mind could have arranged for any message so flawlessly coherent, so strangely imaginative, and sometimes downright bizarre. It must be the work of an Alien Intelligence! ... Whether God is the message, wrote the message, or whether it wrote itself is unimportant in our lives. We can safely drop the traditional idea of the Demiurge, for there is no scientific evidence for a Creator of the natural world, no evidence for a will or purpose that goes beyond the known laws of nature. Even the evidence of life on earth, which promoted the compelling 'argument from design' for a Creator, can be accounted for by* *evolution... So we have a message without a sender"* [Pag88]. *"The feature of living matter that most demands explanation is that it is almost unimaginably complicated in directions that convey a powerful illusion of deliberate design"* [Daw01]. *"A Designer is a natural, appealing and altogether human explanation of the biological world. But as Darwin and Wallace showed, there is another way, equally appealing, equally human, and far more compelling: natural selection, which makes the music of life more beautiful as the aeons pass"* [Sag60]. *"The kind of explanation we come up with must not contradict the laws of physics. Indeed it will make use of the laws of physics, and nothing more than the laws of physics"* [Daw96Bp15]. *"Natural selection is the blind watchmaker, blind because it does not see ahead, does not plan consequences, has no purpose in view. Yet the living results of natural selection overwhelmingly impress us with the appearance of design as if by a master watchmaker, impress us with the illusion of design and planning. The purpose of this book is to resolve this paradox to the satisfaction of the reader, and the purpose of this chapter is further to impress the reader with the power of the illusion of design"* [Daw96Bcover].

"Darwin's theory is now supported by all the available relevant evidence, and its truth is not doubted by any serious modern biologist. But, important as evidence is, in this article I want to explore the possibility of developing a different kind of argument. I suspect that it may be possible to show that, regardless of evidence, Darwinian natural selection is the only force we know that could, in principle, do the job of explaining the existence of organised and adaptive complexity" [Daw82]. *"Darwinism is the only known theory that is in principle capable of explaining certain aspects of life... even if there were no actual evidence in favour of the Darwinian theory"* [Daw96Bp287-288]. *"The theory of evolution by cumulative natural selection is the only theory we know of that is in principle capable of explaining the existence of organized complexity. Even if the evidence did not favour it, it would still be the best theory available!"* [Daw96Bp317] Faith ignores evidence!

To expound some revolutionary views, Paul Davis (professor of theoretical physics) writes: *"The really amazing thing is not that life on Earth is balanced on a knife-edge, but that the entire universe is balanced on a knife-edge, and would be total chaos if any of the natural 'constants' were off even slightly... even if you dismiss man as a chance happening, the fact remains that the universe seems unreasonably suited to the existence of life -- almost contrived -- you might say a 'put-up job'"* [DavWiki]. Philip Skell, chemist and member of the United States National Academy of Sciences, writes: *"Darwinian evolution – whatever its other virtues – does not provide a fruitful heuristic in experimental biology... The claim that it is the cornerstone of modern experimental biology will be met with quiet skepticism from a growing number of scientists in fields where theories actually do serve as cornerstones for tangible breakthroughs"* [Ske05]. *"Yet, under ordinary conditions, no complex organic molecule can ever form spontaneously, but will rather disintegrate, in agreement with the second law. Indeed, the more complex it is, the more unstable it will be, and the more assured, sooner or later, its disintegration. Photosynthesis and all life processes, and even life itself, cannot yet be understood in terms of thermodynamics or any other exact science, despite the use of confused or deliberately confusing language"* [Str77]. *"Once we see that life is cosmic it is sensible to suppose that intelligence is cosmic. Now problems of order, such as the sequences of amino acids in the chains which constitute the enzymes and other proteins, are precisely the problems that become easy once a directed intelligence enters the picture... So if one proceeds directly and straightforwardly in this matter, without being deflected by a fear of incurring the wrath of scientific opinion, one arrives at the conclusion that biomaterials with their amazing measure or order must be the outcome of intelligent design. No other possibility I have been able to think of in pondering this issue over quite a long time seems to me to have anything like as high a possibility of being true"* [Hoy82]. *"The Darwinian theory is wrong and the continued adherence to it is an impediment to discovering the correct evolutionary theory"* [Hoy99].

"The failure of purely physical theories to describe or explain information reflects Shannon's concept of entropy and his measure of 'news.' Information is defined by its independence from physical determination: If it is determined, it is predictable and thus by definition not information. Yet Darwinian science seemed to be reducing all nature to material causes" [Gil06]. *"Our experience-based knowledge of information-flow confirms that systems with large amounts of specified complexity (especially codes and languages) invariably originate from an intelligent source"* [Mey04]. *"Biologists must be encouraged to think about the weaknesses of the interpretations and extrapolations that theoreticians put forward or lay down as established truths. The deceit is sometimes unconscious, but not always, since some people, owing to their sectarianism, purposely overlook reality and refuse to acknowledge the inadequacies and the falsity of their beliefs"* [Gra77p8]. *"There have been an awful lot of stories, some more imaginative than others, about what the nature*

of that history [of life] *really is. The most famous example... is the exhibition on horse evolution prepared perhaps fifty years ago. That has been presented as the literal truth in text-book after text-book. Now I think that this is lamentable, particularly when people who propose those kinds of stories may themselves be aware of the speculative nature of some of that stuff"* [Eld85]. *"It is ironic that the scientific facts throw Darwin out, but leave William Paley, a figure of fun to the scientific world for more than a century, still in the tournament with a chance of being the ultimate winner... Indeed, such a theory is so obvious that one wonders why it is not widely accepted as being self-evident. The reasons are psychological rather than scientific"* [Hoy81Ep130].

"Much of the vast neo-Darwinian literature is distressingly uncritical... Natural selection has shown insidious imperialistic tendencies" [Fod10S]. *"There is something wrong – quite possibly fatally wrong – with the theory of natural selection ... Neo-Darwinism is taken as axiomatic; it goes literally unquestioned... Natural selection can't be the mechanism of evolution... Introducing mental states into the operation of natural selection would allow it to reconstruct the distinction between selection and selection-for... but the cost would be catastrophic. Mental processes require minds in which to happen"* [Fod10Wpviii-ix,114&155]. "*It is relatively well known how organisms adapt to their environment... However, whether this knowledge suffices to explain macroevolution, narrowly defined here to describe evolutionary processes that bring about fundamental novelties or changes in body plans has remained highly controversial"* [Hin06]. *"The gene as the sole agent of variation and unit of inheritance, and the dogmatic insistence on this stance by the popularizers of the Synthesis* [neo-Darwinism], *quelled all calls for more comprehensive attitudes. Although gene centrism has been a major point of contention, including strong criticism from philosophy of science, this aspect could not be changed from within the paradigm of the MS, which rested on it both explicitly and implicitly"* [Pig10].

NASA Astrobiology Institute Chief Bruce Runnegar says, "*Natural selection is not a mechanism, it's the process by which the results of evolution are sorted... all of the processes are much more complicated than people imagine. There are many more loops in the biochemistry of organisms. There are many cases where the RNA itself does the job and feeds back into the protein loop. So this whole system has become so much more complex. We understand the nature of life a lot more than we did 10 years ago"* [Maz10p188&190]. Microbiologist and physicist Carl Woese. *"How could modern biology have gone so badly off track?... it is a simple tale of scientific complacency... with the establishment of the genetic basis of inheritance: Mendel's genetics combined with Darwin's theory of evolution by natural selection. Biologists refer to this as the 'modern synthesis,' and it has been the basis for all subsequent developments in molecular biology and genetics... biologists were seduced by their own success into thinking they had found the final truth about all evolution"* [Woe10].

"A Chinese paleontologist lectures around the world saying that recent fossil finds in his country are inconsistent with the Darwinian theory of evolution... When this conclusion upsets American scientists, he wryly comments: 'In China we can criticize Darwin but not the government. In America you can criticize the government but not Darwin.'... one reason the science educators panic at the first sign of public rebellion is that they fear exposure of the implicit religious content in what they are teaching" [Joh99]. Presidential Medal of Science winner Lynn Margulis notes, *"as far as 'survival of the fittest' goes, ...* [natural selection is] *neither the source of heritable novelty nor the entire evolutionary process...* [making] *Darwinism 'dead,' since there's no adequate evidence in the literature that random mutations result in new species... Natural selection is the failure to reach the potential, the maximum number of offspring that, in principle, can be produced by members of the specific species in question"* [Maz10p257&267]. Michael Polanyi turned to philosophy at the height of his scientific career when he saw how ideologies were being employed to hinder free scientific expression and inquiry. Polanyi argued life is not reducible to physical and chemical principle, but rather, *"the information content of a biological whole exceeds that of the sum of its parts"* [PolWeb]. The argument for abiogenesis *"simply says it happened. As such, it is nothing more than blind belief. Science must provide rational theoretical mechanism, empirical support, prediction fulfillment, or some combination of these three. If none of these three are available, science should reconsider that molecular evolution of genetic cybernetics is a proven fact and press forward with new research approaches which are not obvious at this time"* [Tre04].

Intelligent people interpret data differently, largely based on presuppositions and training. Currently, those espousing the undirected natural processes scenarios are in control of the vast majority of scientific clout. Those holding the view that the evidence shows empirically detectable intelligent design are largely shunned as being "unscientific." Studies have repeatedly shown that the public holds views that are more compatible with ID, yet undirected naturalism is taught as "truth" in public school systems, despite its lack of scientific evidence. Since Atheism, Secular Humanism, and pure naturalism (to which all is subservient) have been confirmed as religions by US courts, and Evolution has been declared a religion by evolutionists, this teaching may be against the First Amendment religious establishment clause. Those who want to promulgate undirected naturalism as unassailable dogma certainly have the right to establish private schools (not tax supported) to carry out that objective. Students currently are being taught half-truths and even falsehoods as if they were true.

There are three topics that should not be included in science unless/until sufficient facts from known science demonstrate feasible scenarios: the origin of the mass and energy of the Universe, the origin of life, and the origin of species. To pretend that a "scientific-sounding" scenario is actually science in these areas does a disservice to both science and the public, and diminishes the reputation of all science by anyone examining the evidence. It would be appropriate to teach why those areas are not science, and that science has no scientific answers to questions of origins. When evolution is taught, its scope should be limited to that which is known, eliminating unwarranted speculation. Evolution has been a knowledge-stopper instead of an enlightener. For example, how long did it take to actually study the vestigial organs to discover that all 180 that were considered useless remnants of evolution are actually useful or even vital? More recently, what for over 20 years was considered "junk DNA" caused by mutations which haven't yet mutated into useful genes, has found many useful purposes, and seems likely to yield more positive results. When ID is taught, only the empirical scientific evidence (such as irreducible complexity or functional and prescriptive information) should be taught, with the fact that science cannot know who or what caused the design (obviously any theological implications should not be addressed in a public school). Discovery Institute has information on the Kitzmiller decision [Dis07] and information for educators [Wel07] interested in learning more.

When reading or hearing findings that purport to be science, be critical. This applies to peer reviewers, students, and the general public (e.g. TV's "science programs"). If "possible" is used, analyze whether or not non-zero probability has been proven using known science. If it hasn't, don't treat the scenario as science, but rather as faith or speculation. A good benchmark for taking seriously any abiogenesis scenario would be if that scenario were awarded the $1 million Origin of Life Prize [OOLprize]. If that prize is awarded, the winning scenario may be considered science. Until then, view any abiogenesis scenario with extreme skepticism. No scenario should be published as science if it fails the feasibility test described on page 3 [Abe09U].

8 Why Intelligent Design?

The term "intelligent design" was perhaps first used in 1897 by Oxford's F. Schiller: *"it will not be possible to rule out the supposition that the process of Evolution may be guided by an intelligent design"* [Sch1897]. *"The theory of intelligent design holds that certain features of the universe and of living things are best explained by an intelligent cause, not an undirected process such as natural selection... The scientific method is commonly described as a four-step process involving observations, hypothesis, experiments, and conclusion. Intelli-gent design begins with the observation that intelligent agents produce complex and specified information (CSI). Design theorists hypothesize that if a natural object was designed, it will contain high levels of CSI. Scientists then perform experimental tests upon natural objects to determine if they contain complex and specified information. One easily testable form of CSI is irreducible complexity, which can be discovered by experimentally reverse-engineering biological structures to see if they require all of their parts to function. When ID researchers find irreducible complexity in biology, they conclude that such structures were designed.* [Idweb]. *"Irreducible complexity is indeed a valid criterion to distinguish between intelligent and nonintelligent design"* [Pig04].

"For an irreducibly complex system, each of the parts of the irreducible core plays an indispensable role in achieving the system's basic function. Thus, removing parts, even a single part, from the irreducible core results in complete loss of the system's basic function... To determine whether a system is irreducibly complex therefore employs two approaches: (1) An empirical analysis of the system that by removing parts (individually and in groups) and then by rearranging and adapting remaining parts determines whether the basic function can be recovered among those remaining parts. (2) A conceptual analysis of the system, and specifically of those parts whose removal renders the basic function unrecoverable, to demonstrate that no system with (substantially) fewer parts exhibits the basic function. Indispensable parts identified in step (1) and then confirmed in step (2) to admit no simplification belong to the irreducible core of an irreducibly complex system" [Dem04p4-5].

"With a bacterial flagellum, you're talking about a machine that's got forty structural parts. Yes, we find ten of them are involved in another molecular machine, but the other thirty are unique. So where are you going to borrow them from? Eventually you're going to have to account for the function of every single part as if

originally having some other purpose. I mean you can only follow the argument so far, until you run into the problem that you're borrowing from nothing" [MinVid]. Changing the original 10^{-1170} probability for the spontaneous formation of the flagellum for the remaining two-thirds of flagellar proteins, yields something like 10^{-780} [Dem04], which is still clearly infeasible. The problem Behe *"has raised is, if anything, still more vexing for Darwinism that when he first raised... even when the most generous allowance of legitimate advantages, the probabilities computed for the Darwinian mechanism to evolve irreducibly complex biochemical systems like the bacterial flagellum always end up being exceedingly small"* [Dem04]. *"Biological functionality is turning out to be much more highly specified and precise than we had originally envisioned... biology is really a science of engineering, where the constraints for bio-functionality are extreme – to the point that nearly every molecular interaction is remarkably precise and tightly controlled. Molecular biology is much like a jigsaw puzzle where each piece must be specifically shaped to fit with the other pieces around it"* [Bra03].

"Our experience with information-intensive systems (especially codes and languages) indicates that such systems always come from an intelligent source--i.e., from mental or personal agents, not chance or material necessity. This generalization about the cause of information has, ironically, received confirmation from origin-of-life research itself. During the last forty years, every naturalistic model proposed has failed to explain the origin of information--the great stumbling block for materialistic scenarios. Thus, mind or intelligence or what philosophers call 'agent causation' now stands as the only cause known to be capable of creating an information-rich system, including the coding regions of DNA, functional proteins, and the cell as a whole. Because mind or intelligent design is a necessary cause of an informative system, one can detect the past action of an intelligent cause from the presence of an information-intensive effect, even if the cause itself cannot be directly observed... Because we know intelligent agents can (and do) produce complex and functionally specified sequences of symbols and arrangements of matter (i.e., information content), intelligent agency qualifies as a sufficient causal explanation for the origin of this effect. Since, in addition, naturalistic scenarios have proven universally inadequate for explaining the origin of information content, mind or creative intelligence now stands as the best and only entity with the causal power to produce this feature of living systems" [Mey00].

The designer's identity cannot be determined by empirical science, and therefore is not part of ID. *"It is important to stress that the refusal of ID proponents to draw scientific conclusions about the nature or identity of the designer is principled rather than merely rhetorical. ID is primarily a historical science, meaning it uses principles of uniformitarianism to study present-day causes and then applies them to the historical record in order to infer the best explanation for the origin of the natural phenomena being studied. ID starts with observations from 'uniform sensory experience' showing the effects of intelligence in the natural world... Scientists have uniform sensory experience with intelligent causes (i.e. humans), thus making intelligence an appropriate explanatory cause within historical scientific fields. However, the 'supernatural' cannot be observed, and thus historical scientists applying uniformitarian reasoning cannot appeal to the supernatural. If the intelligence responsible for life was supernatural, science could only infer the prior action of intelligence, but could not determine whether the intelligence was supernatural"* [DeW07]. For example, that this book was designed can easily be verified by science, which would be true if the author were born as a normal human, transplanted from an alien planet, or existed forever. The identity of the designer is a separate issue from the detectability of design. The mechanism of design implementation is also a separate issue; e.g. did this author use a keyboard, a voice recognition device, a secretary, or mental telepathy during the book's creation?

Some, such as palaeobiologist Conway Morris, propose that *"Minds may be not only universal, but also the same"* [Mor07], so that mind (intelligence) is a universal principle. Einstein wrote *"Certain it is that a conviction, akin to religious feeling, of the rationality or intelligibility of the world lies behind all scientific work of a higher order... This firm belief, a belief bound up with deep feeling, in a superior mind that reveals itself in the world of experience, represents my conception of God*" [Ein54]. It seems likely that most people have a belief about the designer(s) based on personal philosophy. That belief is not based on empirical science, and may include a designer that is supernatural, alien, naturally-intrinsic, or non-existent (discounting a designer is also just a belief that cannot be proved).

Detectability of design has been questioned by some people. Dembski [Dem98] illustrates detection of rigged ballot placement (design) by Nicholas Caputo, who 40 out of 41 times placed a Democrat at the top of the ballot. The New Jersey Supreme Court believed the 2×10^{-11} probability of such placement history showed design. A new computer entropy-based security system can detect designed viruses and worms based on Internet traffic patterns [Nuc07]. Watermarks in Venter Institute's designer genes have been detected [Mad08], demonstrating that ID can be detected in biological systems. Reverse engineering of electronics or computer programs routinely is done to determine the intelligently designed functions. Reverse engineering the complex computing systems of life to determine characteristics of the native language(s) and the operating system(s) of the enzymes may yield breakthroughs in miniaturization and performance. If intelligence couldn't be detected by attempting to extract information from data, the SETI program is worse than useless (since billions have been spent on it). Forensic science routinely determines whether a death is natural, accidental, or designed (murder). For example, suppose a body was found in a street, run over by a car. One might assume the death was accidental, but what if paint on the body was traced to an ex-spouse's car following a bitter divorce? Or what if an autopsy determined the person died of a heart attack? The data would be used to infer the cause of the historic event.

"You cannot define intelligence; trying to do so only leads toward regress. One must try to design intelligence from the ground up, letting the agent's actions in its environment speak for themselves" [Shr98]. Intelligence is detected by its effects. It is detectable in a number of ways, but usually is detected by effects that defy the assumption of chance. Dawkins writes: *"The more statistically improbable a thing is, the less can we believe that it just happened by blind chance. Superficially the obvious alternative to chance is an intelligent Designer. But Charles Darwin showed how it is possible for blind physical forces to mimic the effects of conscious design, and, by operating as a cumulative filter of chance variations, to lead eventual*[ly] *to organized and adaptive complexity, to mosquitoes and mammoths, to humans and therefore, indirectly, to books and computers*" [Daw82]. In other words, those espousing undirected natural causes for everything, believe that the original chance formation of the Universe led to chance formation of life which led to humans which led to computers, etc. Tracing the propositional logic backwards leads to chance causing the computer that is being used to record the thoughts (also caused ultimately by chance) for this book. Ignoring the fact that chance has no "causative effect," where is the proof of such a belief? What if in: *"Biology is the study of complicated things that give the appearance of having been designed for a purpose"* [Daw96Bp1], the reason for the appearance is that they **are** designed? What if the Universe's constants that *"appear*

to be finely, even heroically, tuned" [AAAS05], appear that way because they **are** fine-tuned?

The limits of science may be philosophical, rather than scientific. *"What are the limits for what can claim to be science? One proposal is methodological naturalism (MN), which requires that scientific theories can postulate only natural causes. What are the limits for what MN-Science can claim to explain? If we decide to accept methodological naturalism, a second limit is logically necessary: If an event really does involve a non-natural cause, any explanation of the event by MN-Science (in terms of only natural causes) will be incomplete or incorrect"* [RusWeb]. *"Limiting science to a predetermined set of acceptable explanations naturally begs the question, 'What if there is no natural explanation?' What if, in fact, an intelligent agent was responsible for DNA, etc.? Science would forever miss it and would continue to squander intellectual and financial capital on finding naturalistic answers that do not exist. Scientific progress depends heavily upon discovering blind alleys and rejecting failed theories. This is simply the way that science works, and thus, ID theory should be seen as invigorating, not stifling, scientific investigation"* [Bow07]. Naturalism has tried to squeeze the data into prevailing suppositions, refusing to even consider alternate scenarios that are at least as scientific, and refusing to fund or publish those alternatives. The idea that positing an intelligent cause somehow is restricting to science has no basis, as can be seen by considering the great scientists of the past (such as Newton, Faraday, Pascal, Kepler, Maxwell, and Boyle).

The benefits of an ID model are potentially wide-ranging. If ID had been accepted, virtually all data recently found concerning the Universe's fine tuning, the complexity of life, and the information of life (including in "junk DNA") would be seen as confirmation of the ID model. Investigation of information processing systems should be able to use existing knowledge and intelligent principles without trying to speculate about how such complex interacting systems could arise via undirected nature. Perhaps such research could determine the "native" machine languages of life's processors, and where the operating systems reside (in life's hardware or software or some combination) and the fundamental computing operations involved. Whereas some dismiss the repetitive nature of some non-coding DNA as unimportant, this author's experience with repetitive computer programming for supercomputing (to minimize overhead of context switching) and for magnetic tape operating systems (serial, rather than random memory access), highlights the importance of repetitive structures. In the case of DNA as memory to be read by enzyme "computers," the access would necessarily be sequential so that identical "machine language" instructions in the DNA would require repeating. It may be that investigation of those repeated structures will yield insight into the native machine language of life. The idea that non-coding DNA is "junk" makes as much sense as viewing the instructions in a computer program as unimportant since all that is important are the encoded data. Accepting ID would mean that research would not be restricted to that which is based on undirected natural causes, and lifting those restrictions may yield findings hitherto unimagined. Professor Frederick Crews grudgingly acknowledges that: *"Intelligent design is thriving... in cultural circles where illogic and self-indulgence are usually condemned"* [Cre01].

Occam's razor (a philosophical principle) *"states that the explanation of any phenomenon should make as few assumptions as possible, eliminating those that make no difference in the observable predictions of the explanatory hypothesis or theory... entities must not be multiplied beyond necessity... 'All other things being equal, the simplest solution is the best.' In other words, when multiple competing theories are equal in other respects, the principle recommends selecting the theory that introduces the fewest assumptions and postulates the fewest entities"* [OccWik]. *"The burden is on the skeptic to provide good rational or empirical grounds for believing that the false appearance of design is not unlikely under actual conditions. Design sceptics, in common with other sceptics from antiquity to the modern day, attempt an illegitimate shift in the burden of proof. The skeptic attempts to rebut a successful design inference by merely raising the possibility that the appearance of design may be illusory, challenging the defender of the inference to prove a negative – to prove that the skeptical scenarios could not have happened. The appropriate response to such skeptical challenges is to place the burden of proof where it belongs: the skeptic must provide substantial and specific grounds for doubting the soundness of the design inference in the particular case in question"* [Koo01].

In the absolute sense, one cannot rule out design of anything since a designer could design something to appear as if it weren't designed. For example, one may not be able to prove an ordinary-looking rock hadn't been designed to look as if it were the result of natural processes. The "necessity of design," however, is falsifiable. To do so, merely prove that known natural processes can be demonstrated (as opposed to merely speculated from unknown science) to produce: the fine-tuning empirically detectable in the Universe, life from non-life (including the information and its processing systems), the vast diversity of morphology suddenly appearing in the Cambrian era, and the increasing complexity moving up the tree of life (with the accompanying information increase and irreducibly complex systems). If those can be demonstrated with known science, the "necessity of design" will have been falsified in line with using Occam's Razor principles for determining the most reasonable scenarios. If the "necessity of design" is falsified, some may continue to BELIEVE in design, but ID would no longer be appropriate as science.

We have seen the improbabilities of the physical constants and composition of the earth and Universe. Any other physical artifact having orders of magnitude less fine-tuning would be acknowledged as being designed without debate. We have seen the requirements that known science requires all components, including DNA, RNA, enzymes, and other proteins be present for life, and that life cannot be determined just by chemistry and physics. We have seen the vast functional information content of even the simplest life, and that it is impossible for life to have arisen "proteins first." We know of no case in which prescriptive or complex functional information, even considerably lower than that in DNA, has arisen without the actions of an intelligent agent. We have considerable experience with studying the effect of humans as intelligent agents. We know that mind can direct the production of extremely complex artifacts, such as this book or computer programs, with planning and future goals in mind. We have examined the prescriptive, functional, and Shannon information of life, with Shannon information placing limits on information transfer, including the channel capacity limit that requires an initial alphabet at least as complex as the DNA codon alphabet. We have seen irreducibly complex biological artifacts that require all parts to simultaneously appear in an organism. We have looked at the Cambrian explosion to see a sudden influx of specified complexity for the formation of features in most (if not all) phyla, with no predecessor fossils. We have examined the recent findings of "junk DNA," which ID proponents would have encouraged examination over 20 years sooner, rather than dismissing it as evidence for Darwinian evolution.

To recap the arguments for ID, trying to put realistic values on the probability that ID is true, we start with the fine-tuning of the Universe, since origin of mass and energy is scientifically unknowable. Just three numbers (ignoring the precision of original entropy of $_{10}$-10^{123} [Pen89]): star formation probability of 10^{-229} [Smo97], cosmological constant precision of 10^{-120} [Mic99], and the necessary 10^{500} "universes" to make feasible the evident fine-tuning [AAAS05], show that design is many orders of magnitude more than 10^{500} times as likely than undirected processes. Since either completely undirected or directed (including "partially directed") processes "caused" the fine-tuning, the sum of those two probabilities must equal 1. Therefore, the probability of ID is greater than $(1 - 10^{-500})$, or 0.999... (at least 496 more 9s), nearly certain. For abiogenesis, we could consider 10^{-175} as the probability of forming a simple life-compatible protein [Tha92p156-157] or the simplest form of living organism known as $10^{-340,000,000}$ [Mor79]. There really is no need for those, however, since the probability of an undirected source of information contained in life is 0 (impossible) based on life's choice-contingency instantiated algorithms, functional information (including semiotic systems), and alphabet requirements for information transfer, making the probability of ID exactly 1 for the first life. It is infeasible to put a reliable estimate on the probability of undirected production of the tree of life by undirected causes (considering no new net information results by mutations, irreducible complexity, Cambrian explosion, etc.). Since the functional information of human DNA is over 10^7 bits more than the simplest organism, somehow over 10^7 bits would have to be injected into the genome, but there's no known mechanism for that to happen. Using functional information, the ratio of formation probabilities of human to simplest organism has been shown to be less than $10^{-3000000}$. Since the result pointing to ID is already a certainty according to known science, the conclusion is that ID has a probability of one (P = 1).

It should be noted that all science is tentative, so new findings may require modification of the above conclusion which is based on today's known science. At this point, one cannot say "it's possible new findings may cause rejection of ID," since that would assume such findings are possible (non-zero probability), which has not been proven by known science. Those disagreeing with this conclusion are invited to show from known science that it is incorrect. If that demonstration cannot be made, one must realize that continued rejection of ID is based on philosophy, not science.

Appendix A: Math Basics – Exponents, Logarithms, and Probability

If the math concepts of this book are unfamiliar, this Appendix may require proceeding very slowly, until each item is understood. It will clarify the scientific meaning of terms like "possible," "impossible," and "probable." For example, a weather forecaster statement "it will probably rain tomorrow," and it doesn't rain, isn't an incorrect prediction, unless rain's probability were actually less than 50%.

Scientific or exponential notation is convenient when expressing the very large or very small numbers used in this book. Examples include:

$4000000000 = 4$ billion $= 4 \times 10^9$ (count digits to the right of the first digit)

$0.000001 = 1$ millionth $= 1/10^6 = 10^{-6}$

(count to the right of the "." including the first non-zero digit)

$602,214,179,000,000,000,000,000 = 6.02 \times 10^{23}$ (Avagadro's number)

Avagadro's number is the number of hydrogen atoms in one gram or the number of carbon atoms in 12 grams. A 1 carat diamond contains approximately 10^{22} carbon atoms. A diamond containing 10 trillion (10^{13}) atoms would be a billionth (10^{-9}) carat diamond. At 4800 characters per page, 6.02×10^{23} characters would require a stack of pages about 77,000 times the distance to the sun. Notice that each unit change in exponent is a factor of ten more or less:

$10^9 = 10 \times 10^8 = 10^{10}/10 = 1000 \times 10^6 = 1/10^{-9} = 10/10^{-8}$ (all have the same value).

The Richter scale for expressing earthquake magnitude is based exponentially, so that a magnitude 5 earthquake is 100 times as strong as a magnitude 3 and 100 times weaker than a magnitude 7 quake. Since 6 is 20% higher than 5, if one is thinking linearly, rather than exponentially, one may visualize a magnitude 6 quake as only 20% stronger than a magnitude 5 quake, instead of the 900% stronger (10 times) that it really is. Numbers expressed exponentially may appear to be considerably different than what they represent. For example, a googol is 10^{100}, but is physically a totally hypothetical number since the maximum estimate (many are much less) of number of atoms in the Universe is 10^{80} [Sag79]. There is not a googol of anything physical in the Universe.

Logarithms can be used to calculate the exponent for any base: $\log_b(b^x) = x$. Note, $b^x \times b^y = b^{x+y}$ and $\log_b(b^x \times b^y) = x + y$. e.g.: $\log_{10}(1000) = 3$, $\log_{10}(1/1000) = -3$, $\log_{10}(5) \sim 0.6990$, $\log_{10}(50) \sim 1.6990$, $\log_{10}(.5) \sim -0.3010$ ($= 0.6990 - 1$), $\log_2(8) = 3$ ($2^3 = 8$), $\log_2(.25) = -2$ ($1/4 = 2^{-2}$), $\log_{10}(6.02 \times 10^{23}) = \log_{10}(10^{23.780}) = 23.780$, $\log_{10}(1) = \log_2(1) = \log_e(1) = 0$, $\log_e(10) \sim 2.3025$; $\log_e$ is the natural logarithm, where

$e = -\sum_{i=0}^{\infty} 1/i! = 1/1 + 1/2 + 1/6 + 1/24 + 1/5! + \ldots$ ($\sum$ means to sum all terms, ! is factorial.)

The law of probability expresses the likelihood of a particular outcome from within the set of possible outcomes. Probability has a range of 0 (impossible) to 1 (certain). Rolling a die has a probability of 1/6 (i. e.1-in-6) for any particular number. Rolling a five 10 times in a row results in a 1/6 probability of a five on the next die roll since chance has no causative effect. Some people who don't understand chance play the lottery by betting on their lucky numbers, betting on the numbers that have been chosen most frequently, or betting on numbers that have been chosen least frequently (reasoning that those numbers would have to be chosen more frequently in order to catch-up with their a priori probability). For example, in the California SuperLotto lottery, one picks five different numbers from 1 to 47 and one MEGA number from 1 to 27. If those numbers match the numbers drawn by the Lottery, a win results (this discussion will be limited to a jackpot win, not a partial win by matching some numbers). Note that the order of numbers chosen makes no difference. The first choice has a probability of 5/47 of being correct since there are 5 correct numbers in the 47 possibilities. If the first number is correct, the second number chosen has a probability of 4/46 since there are only 4 correct numbers out of the 46 numbers remaining. Similarity, the third through fifth choices have probabilities of being correct of 3/45, 2/44, and 1/43, respectively. The final "MEGA" number has a probability of 1/27 since the possibilities are 1 to 27.

The probability of independent consecutive events is the product of the probabilities of the individual events. The law of probability calculates that any arbitrary choice of numbers (regardless of how chosen) has a probability of winning the jackpot of one in 41,416,353, with the product of all six correct probabilities being 2.4×10^{-8} (which can also be expressed using combination factorial notation as 5!42!/47!/27, where n! is the product of all integers up to and including n). Since the probability of winning plus the probability of not winning is 1 (a certainty), the probability of failing to win (P_f) the jackpot is 0.999999976 ($= 1 - 2.4 \times 10^{-8}$), a figure you won't see in Lottery promotions. The probability of n consecutive failures is P_f^n. To compute the number of trials needed to make a win as likely as continued consecutive failures, solve the equation $0.5 = P_f^n$, or $n = \log_2(0.5)/\log_2(P_f) = -1/\log_2(P_f) = 2.9 \times 10^7$. A Lottery player using one card each Wednesday and Saturday would probably have a win only after 2,767 centuries. If a winner buys a single ticket for the next lottery, the chance of winning the second time is 2.4×10^{-8}. Note, however, that the probability of winning twice in a row is $(2.4 \times 10^{-8})^2$, or 5.8×10^{-16}. This is obviously very unlikely to happen. A first win results in people attributing the win to chance luck, whereas additional wins result in people attributing the wins to a rigged contest (design) because of the extreme unlikelihood of such an outcome by chance.

The probability of selecting a particular atom of the Universe randomly is 10^{-80}. The probability of rolling ten fives in a row is $(1/6)^{10} = 1/60466176 = 1.6 \times 10^{-8}$. The probability of tossing 20 consecutive heads is 0.5^{20} or 9.53×10^{-7}. The probability of being dealt a royal flush from a normal shuffled deck of 52 cards is $(20/52) \times (4/51) \times (3/50) \times (2/49) \times (1/48) = 1/649740 = 1.54 \times 10^{-6}$. Assume the 48 contiguous US states were covered with densely-packed [Ste99] trees, each with 200,000 leaves [Wi-Web]. If one leaf were marked, and all leaves mixed together, the probability of choosing the marked leaf would be 3.1×10^{-17}. If the 48 states were covered 15 feet deep in cents and a single 1943 copper cent tossed into the pile, the probability of choosing the 1943 coin would be 10^{-18}.

Note that increasing the number of attempts for something that is impossible will not increase its likelihood since the probability of failure is $(1\text{-}0)^n = 1$ for any n, so the probability of success is still 0. Likewise, the probability of a certainty doesn't decrease by having more trials since the probability of failure is $(1\text{-}1)^n = 0$ and 1-0 is still 1 (certain). If the probability isn't 0, theoretically it could happen. For example, the probability of throwing 300 consecutive heads is 0.5^{300} or 4.91×10^{-91}. Since chance has no causative effect, the desired pattern could happen on the first attempt, but it is less likely than winning the Lottery 12 consecutive times.

Sometimes the probability of an outcome is uncertain. For example, if a coin being tossed may bounce differently when landing heads or tails, each may have a probability different than 0.5. Is it possible (non-zero probability) for a tossed die to end up on-edge? Just because an outcome hasn't yet been observed doesn't necessarily mean it couldn't happen. If a detailed analysis of cube motions determined an on-edge result was possible, then it may be desirable to include it in the probability expression. In science, "possible" means a non-zero probability, which should only be used when known science demonstrates that to be true. It would not be scientifically accurate to state "it is possible that the die may end up on-edge, reducing the 1/6 probability of each numeric result" unless one first demonstrates with known science that the on-edge result is possible. In case you're wondering, there have been reported instances of coins ending up on-edge [Mur93], and also coin tosses having biased probability [Dia07].

Appendix B: Origin of Mass and Energy Scenarios

The law of conservation of mass and energy says that matter and energy cannot be created nor destroyed, but may be converted from one form to another ($\Delta E + c^2 \Delta M = 0$, where c is the speed of light). Before Einstein, each change (ΔE or ΔM) was assumed to be 0, but science now knows that mass can be converted to energy, and vise versa. If a gram of mass is converted (ΔM is negative) into energy, such as in an atomic bomb, 9×10^{13} joules of energy are produced. Note that the atomic bomb that leveled Hiroshima was over 160 times less than this with a 6.4 mg loss of nuclear mass. In a nuclear power plant, 2700 MWHrs of electric power is produced per gram of mass converted. If one could convert the mass of a can of soda into energy, it would produce enough electric power for a million homes in the United States for a year. When one considers the 3×10^{69} joules of energy required to produce the estimated 3×10^{55} g mass of the Universe [NASA], one is forced to look beyond the random energy fluctuations that were once proposed as a source. *"The principle of the conservation of energy is considered to be the single most important and fundamental 'law of nature' known to science, and is one of the most firmly established. Endless studies and experiments have confirmed its validity over and over again under a multitude of different conditions"* [You85].

Since undirected natural processes producing "matter and energy from nothing" has been dismissed by virtually all scientists (some do believe there is no such thing as "nothing," however), "eternal existence" of matter and energy has been proposed. This model proposes an oscillating Universe that has always existed, alternating between a Big Bang and a Big Crunch. Although this model complies with the conservation law, there are two major problems. One problem is that the known mass of the Universe is insufficient to cause a collapse since the Universe is expanding [Hub29] at a rate faster than the escape velocity (gravitational forces cannot slow the expansion enough to cause collapse). Some have speculated the existence of "dark matter," which can't be seen, but whose mass would make a collapse possible, that is the Universe's density exceeds the Friedmann critical value. Dark matter estimates have shown densities of less than half this required Friedmann critical value. Recent findings indicate that even the rate of expansion is increasing [Pee03], which means the Universe cannot collapse. "Dark energy" is the proposed "explanation" of this increase in expansion rate. What *"is driving this apparently anti-gravitational behavior on the part of the Universe, nobody claims to understand why it is happening, or its implications"* [Ove08].

The Second Law of thermodynamics says that in a closed system, entropy always increases. This is the second difficult scientific problem to accommodate for a Universe of infinite age. Entropy is a measure of randomness (disorganization) or inability to do work (limiting energy transfer between entities). Isaac Asimov has said *"As far as we know, all changes are in the direction of increasing entropy, of increasing disorder, of increasing randomness, of running down. Yet the universe was once in a position from which it could run down for trillions of years. How did it get into that position?"* [Asi73] An infinitely old Universe that is not at maximum entropy violates the Second Law because energy is still being transferred within the Universe, which is a closed system by its "uni-" definition. Some have argued that the Universe isn't closed, since it is expanding without limits. "Closed" has nothing to do with physical size, but rather with matter/energy content. In addition, it has been proven that in an open system *"entropy cannot decrease faster than it is exported through the* [open] *boundary, because the boundary integral there represents the rate that entropy is exported across the boundary"* [Sew05]. Note that directed energy can cause a local entropy decrease without violating the Second Law, since the system encompassing that local system would have an entropy increase. An infinitely old Universe would be energy-dead, with no capacity for work, since one result of the Second Law is that perpetual-motion machines are impossible (zero probability). There would be no available energy when randomness is maximum (after infinite time).

Einstein's general theory of relativity (GTR) has become the basis for several proposed speculations. The main set (for each tensor u and v) of equations relating the curvature tensor (R_{uv}), the spacetime tensor (T_{uv}) is given by the simplified partial differential equation $\mathbf{R_{uv} - (1/2)\, g_{uv}\, R = (8\pi G/c^4)\, T_{uv}}$, ($g_{uv}$ includes first and second derivatives), where c is the speed of light and G is the gravitational constant. Spacetime is an invisible flowing stream that bends in response to objects in its path, as it carries everything in the Universe along its twists and turns. Smaller objects travel through space that is warped by the larger object. Since a second-order differential equation isn't limited to a particular solution, there are, with a variety of assumed starting points, *"an infinite number of solutions. The solution that is used within the scientific community is that given by Einstein. He chose the mathematically simplest equation that relates matter and energy to the curvature of space-time. But in truth, there is no reason why reality should conform to our desire for mathematical simplicity"* [Ast05]. Some of the solutions have resulted in speculation about many esoteric concepts [Mor88].

GTR is widely accepted because it predicts many experimentally verified results such as black holes [Kor95], gravitational bending of light [Edd20], and gravitational time dilation (slows with increasing field strength) [Pou65]. A black hole has that name since even light cannot escape its strong gravitational field as it sucks into itself the surrounding spacetime. Scientists have recently created an artificial (not a real one) black hole in the lab that *"will allow astrophysicists to test predictions made by theorists. Physicists would particularly like to test new theories such as quantum gravity, which seeks to reconcile Einstein's theory of general relativity with quantum mechanics"* [Das08]. Some GTR-allowed possibilities are hypothetical. A wormhole is a hypothetical shortcut through spacetime, allowing apparent faster-than-light travel. A white hole is a hypothetical time reversal of a black hole, ejecting matter from its event horizons. A singularity is a hypothetical place with infinite gravitational field where the curvature of spacetime and the density of matter become infinite.

That GTR seems to produce verifiable results with some starting assumptions does not mean that all possible starting points are indeed "possible" (with non-zero probability in reality). For example, Stenger treats negative time as a GTR-allowed possibility – *"the equations of cosmology that describe the early universe apply equally for the other side of the time axis"* [Ste07p126] -- despite almost universal acceptance that time before Planck time (5.4×10^{-44} sec, the theoretically smallest measurable time unit in quantum mechanics) is unknowable. He and others assume that *"there is no such thing as nothing"* since GTR allows for "empty" space to have "vacuum energy," also known as "ether," which could spontaneously produce (without cause, via quantum tunneling) the mass/energy and the negative gravitational energy of the Universe via spontaneous symmetry breaking. In quantum theory, quanta can exhibit both particle and wave properties. When confronting an energy barrier too high to ascend normally, a quantum can be described by a wave-function that has probabilities of positioning it on either side of the barrier. Quantum tunneling results if the function describing the particle indicates it is on the other side of the barrier.

Perhaps a reality check is in order. Eternal existence of such space would require that it would be at maximum entropy. Speculation [Ste07] that this maximum entropy system gave rise to the physical Universe that is not at maximum entropy violates known science (a maximum entropy system has no available energy). Scientific-sounding statements like *"Since 'nothing' is as simple as it gets, we cannot expect it to be very stable. It would likely undergo a spontaneous phase transition to something more stable, like a universe containing matter"* [Ste07p133], and *"the three great conservation laws are not part of any structure. Rather they follow from the very lack of structure"* [Ste07p131], and constants *"must have changed very rapidly during the first moments of the big bang"* [Ste07p147] are foreign to known science. It should be noted that the largest spontaneous energy equivalent inferred by experiment is sub-atomic (and never on a vacuum with verifiably zero energy input, since even the solar neutrino flux is estimated to be $6 \times 10^{10}/cm^2/sec$ [Cosmic]).

One could speculate that the probability, p, of a free quantum of spontaneous energy, q, is related to time (t) such that $p = e^{-\kappa q^2/t}$, where κ is a constant. In this speculation, higher energy quanta become less likely, related to generation of the quantum and its probability of escaping the anti-quantum (assuming attraction is gravity-like). Assuming non-zero probability for spontaneous generation of quanta of electron (β) energy equivalent and Universe (u) energy equivalent, $p_\beta = e^{-\kappa\beta^2/t}$ and $p_u = e^{-\kappa u^2/t} = e^{-\kappa\beta^2(u^2/\beta^2)/t} = e^{-(\kappa\beta^2/t)r}$, where r is the square of the energy ratio of the Universe to an electron: $r = (u/\beta)^2 = (10^{74})^2 = 10^{148}$. This results in $p_u = p_\beta^{1000(140\ zeros)0000}$. It is unknown the probability of generation or what fraction escapes for a spontaneous very low energy free quantum (β) is since one has never been unambiguously observed. The probability of a free quantum with the energy content of the Universe (u) may be proportional to that low energy probability raised to the 10^{148} power. Some may reason that since this isn't zero, it's possible. This speculation doesn't prove it's possible, but reasons that if it is possible, it is improbable to the extreme, requiring time that is meaningful only near infinity, at which point entropy would be near maximum so that the energy needed to accomplish the feat would be unavailable. To those objecting to this speculation, the challenge is there to show from known science that it is feasible to generate the Universe via quantum tunneling from an eternal vacuum.

"String Theory" has been the basis of several recently proposed multi-verse models. In these models, there are countless other universes in addition to our Universe (the contradiction of the "uni-" is ignored in these models). The building blocks for all of the universes are strings of vibrating energy in at least 10 dimensions (some models have 12 or more dimensions [Super, F-Theo]). It is speculated that our Universe arose out of a collision between previously existing universes, which arose out of collisions of their predecessors, and so forth. Each of these collisions produced matter, energy, and natural laws that are unique to that universe. The natural laws evolved during and shortly after the collision as the 10+ dimensions collapsed to four dimensions for our Universe. These are faith-based models based on unseeable 10-12 dimensions and innumerable unseeable universes. Nima Arkani-Hamed (Harvard) and others propose over 10^{500} universes because fewer would make the fine-tuning that is evident clearly infeasible [AAAS05]. Others note that "String Theory" is not a scientific theory since it cannot be observed, tested, or falsified. *"Alternative universes, things we can't see because they are beyond our horizons, are in principle unfalsifiable and therefore metaphysical"* [Gef05]. *"The trouble is, proponents have not produced an iota of empirical evidence for strings. That's why University of Toronto physicist Amanda Peet – a proponent – recently called string theory a 'faith-based initiative'"* [Rev05]. *"No part of it has been proven, and no one knows how to prove it"* [Smo07]. *"Because our Universe is, almost by definition, everything we can observe, there are no apparent measurements that would confirm whether we exist within a cosmic landscape of multiple universes, or if ours is the only one. And because we can't falsify the idea, ... it isn't science"* [Bru06]. *"If ... the landscape turns out to be inconsistent ... as things stand we will be in a very awkward position. Without any explanations of nature's fine-tunings we will be hard pressed to answer the ID critics"* [Gef05]. It is clear that string theory is not science, but a philosophical belief.

Other models are largely ignored by cosmologists. John Wheeler *"ponders the question whether we humans actually create the laws by our observations, in the way that a magician creates illusion—that what we observe around us is no more real than what we observe at a magic show"* [Pri06]. In the Steady State Theory of Cosmology, new matter is continuously created as the Universe expands [Hoy93, Hoy95]. Plasma cosmology attributes the development of the visible Universe to interaction of electromagnetic forces on astrophysical plasma [Alf90]. The original ambiplasma was an equal mixture of ionized matter and anti-matter that would naturally separate as annihilation reactions released energy. Long-held models by most theistic religions typically involve "infinite" energy being(s) converting energy to mass or otherwise supernaturally creating the mass and energy of the Universe. These models, like the other origin models, cannot be tested or falsified by known science. Other models will undoubtedly arise, and as they do, each will need careful examination to verify any scientific validity.

Appendix C: Technical Shannon and Functional Information Details

Shannon information can be used to calculate (without thermodynamic considerations) the probability of a N-long sequence of symbols from a finite alphabet as very close to 2^{-NH}, where NH is the total Shannon information entropy, as opposed to the lower n^{-N} that is often assumed. For example, iso-1-cytytochrome c has Shannon information entropy of 371.42 for the 113 residues in it, so the $2^{-371.42}$ ($= 1.55 \times 10^{-112}$) probability for the correct sequence is much higher than the probability of 20^{-113} ($= 9.63 \times 10^{-148}$) if all acid residues were equally probable in each position [Yoc05p85]. Note that the probability is still very low for that and any other protein sequence.

Probability theory's sample space makes a clear distinction between genomic nucleotide sequences and the code between the mRNA alphabet and the amino acid alphabet of the proteome. *"Definition: The set of all elementary events, A, in sample space [Ω, A,* ***p****] to each of which probabilities p_i have been assigned is called a probability sample space of elementary events or simply a sample space. The elements A are called random variables. The set of probabilities p_i form a probability vector p"* [Yoc05p14]. With probability vector, $\mathbf{p}_A$, of the elements of alphabet A in a source probability space [Ω, A, $\mathbf{p}_A$] and probability vector, $\mathbf{p}_B$, of the elements of alphabet B in destination probability space [Ω, B, $\mathbf{p}_B$], a code is a unique mapping (code bijection) of the symbols of alphabet A onto the symbols of alphabet B. Since the Shannon entropy of the DNA/RNA alphabets 6 and the Shannon entropy of the protein alphabet is $\log_2 20$, the two alphabets are not isomorphic, and information transfer is unidirectional. The genetic code employs the most economical use of its nucleotides and is optimal, i.e. it is both instantaneously decodable as an initial property and it has the minimum average code word length. Note that speculation of a simpler code (with fewer amino acids) has many inherent problems including providing instant decodability and avoiding error catastrophe. The DNA-mRNA-protein system is discrete because all symbols in the alphabet are defined, memoryless because there is no dependence on previous symbols transmitted, and unconstrained since any symbol may follow any other symbol. The particular message transmitted can therefore be treated as one member of a stochastic ensemble generated by a stationary Markov process. The output from DNA (as with other information sources) can be treated as a random process (even if it isn't) completely characterized by the probability space [Ω, A, $\mathbf{p}_A$], [Ω, B, $\mathbf{p}_B$].

Shannon's Channel Capacity Theorem [Sha48] for a noisy channel states that a communication system can transmit information as close to the channel capacity as desired with an arbitrarily small error rate by incorporating redundancy in an error correcting code. Conditional entropy is a measure of the effect of noise. Mutual entropy is a mathematical measure of the similarity between any two sequences one wishes to compare. Mutual entropy relates the input (x) and output (y) channels via: $\mathbf{I(B;A) = I(A;B) = H(x) - H(x|y)}$, where the conditional (x_i given y_i received) entropy is $\mathbf{H(x|y) = -\Sigma_{ij} p_j \, p(i|j) \log_2 p(i|j)}$, $\mathbf{p_j = \sum_{i}^{n} p_i p(j|i)}$ (which relates the probability vector, $\mathbf{p}$, elements to those of the conditional probability matrix, $\mathbf{P}$), and $\mathbf{H(x) = -\sum_{i=1}^{n} p_i \log_2(p_i)}$ is the information entropy. The Shannon Channel Capacity is also the maximum mutual entropy. For a transmitting system with fewer symbols in $[\Omega, A, \mathbf{p}_A]$ to pass information to $[\Omega, B, \mathbf{p}_B]$, the maximum mutual entropy would be exceeded. Whatever the source alphabet for the first DNA, the Shannon's Channel Capacity ensures that it was at least as symbolically complex as the current DNA codon alphabet. Speculating earlier alphabets only pushes the source of information back to predecessors of those alphabets, each of which would also require as much mutual entropy, and which ultimately would require intelligence to produce the original alphabet and the information processing system that incorporates that alphabet.

*"**Complex emergent systems** of many interacting components, including complex biological systems, have the potential to perform quantifiable functions. Accordingly, we define 'functional information,' $I(E_x)$, as a measure of system complexity. For a given system and function, x (e.g., a folded RNA sequence that binds to GTP), and degree of function, E_x (e.g., the RNA–GTP binding energy), $I(E_x) = -\log_2[F(E_x)]$, where $F(E_x)$ is the fraction of all possible configurations of the system that possess a degree of function E_x... Function is thus the essence of complex systems. Accordingly, we focus on function in our operational definition of complexity. Therefore, although many previous investigators have explored aspects of biological systems in terms of information..., we adopt a different approach and explore information in terms of the function of a system (including biological systems)... functional information increases with degree of function, from zero for no function (or minimum function) to a maximum value corresponding to the number of bits necessary and sufficient to specify completely any configuration of that system"* [Haz07].

*"**Intelligence Hypothesis*** [as used by Kalinsky]*: an attribute that distinguishes a mind from mindless natural processes is the ability of a mind to produce effects*

requiring significant levels of functional information" [Kal08p2]. He points out that natural selection is credited with discovering the nucleotides to code for the thousands of proteins, each with a stable 3-D structure, including those used in extremely complex molecular machines and molecular computers. Rigor can be introduced into analyzing the process using principles of genetic and evolutionary algorithms, which require fitness functions to avoid a blind (e.g. – random walk) search. A fitness function represents *"the requirements to adapt to. It forms the basis for selection, and thereby it facilitates improvements. More accurately, it defines what improvement means. From the problem-solving perspective, it represents the task to solve in the evolutionary context"* [Eib03p19]. The fitness function must contain at least as much functional information as the desired outcome or the information deficit must be made up for in a blind search. If natural selection is responsible for the origin of a protein-coding gene, the amount of functional information in the natural fitness function (that initially would be in a non-life form for abiogenesis) can be estimated by measuring the functional information required by a given protein. A specified level of functionality can become probable after R trials of a specified function with a single sampling probability, P, when: $0.5 = 1 - (1 - P)^R$, or $P = 1 - (1 - 0.5)^{1/R}$. If only one configuration meets the required function, the functional information that could occur by mindless natural processes is: $I_{nat} = -\log_2[1 - (1 - 0.5)^{1/R}]$. It is possible to determine *"the likelihood that an effect required intelligent design, where the greater the difference between the functional information required for the effect and I_{nat}, the more likely it is that intelligent design was required. This would hold true for SETI, archeology, forensic science, and biological life"* [Kal08p6]. For example, any evolutionary process must somehow (regardless of the mechanism) search a sequence space to locate areas where physics produces stable, 3-D structures (500-900 estimated possibilities).

To set an upper limit on abiogenesis I_{nat}, Kalinski assumed the entire mass of the earth consisted of amino acids for building 100-chain proteins, with the entire set reorganizing once per year over the 500 million years of pre-biotic activity, for a total of 10^{55} attempts (R). This results in I_{nat} having estimated maximum of 185 bits of functional information that could result from mindless pre-biotic processes. The biomass of the earth, 7×10^{13} Kg [BioWik], is a tiny fraction of the earth's mass, so fewer than 10^{39} amino acids would actually have been available. The I_{nat} maximum would be lower if realistic values for the available components were used and thermodynamics were considered (no enzymes present). It would be higher if one assumed multiple functional possibilities or more trials. Using Kalinski's estimations, the relative comparison of probabilities (intelligent/natural) can be calculated from the difference in observed-natural functional information: ratio = $2^{(I-185)}$.

The watermarks that were found [Mad08] embedded the Venter Institute's synthetic genome for M. genitalium are formed by choosing base pairs that translate to 60 single-letter amino acid codes spelling words like: VENTERINSTITVTE and CRAIGVENTER (Note "U" is not an amino acid abbreviation, so "V" is used). The 20 possibilities for each position in each word (with only 1 correct/functional combination) result in $I(E_x) = -\log_2(20^{60}) = 259$ bits of functional information. It is therefore more than $2^{(259-185)} = 10^{22}$ times more probable that intelligence (mind) produced the watermarks than a mindless natural processes, which is known to be true in this case.

The estimated frequency of occurrence of stable, folded functional protein domains is between 10^{-64} to 10^{-77} [Axe04], which corresponds to 213-256 bits of functional information required to code for a stable, folded protein domain. It is at least 10^{19} times more probable for ID to produce a folded functional domain than mindless natural processes. Universal proteins RecA, with 1,553 aligned sequences and 688 bits of information, and SecY, with 469 aligned sequences and 832 bits of information, were used to estimate the functional information of an average 300-amino acid protein at 700 bits. Intelligent direction is *"10^{155} times more probable than mindless natural processes to produce the average protein. Again, if natural selection is invoked to explain the origin of proteins, a fitness function will be necessary that requires intelligent design."* [Kal08p11]

Using the 700 bits per protein with the estimated 382 protein-coding genes that the simplest known life form requires [Gla06] yields an estimated 267,000 bits for the simplest known life. This information indicates it is about *"$10^{80,000}$ times more likely that ID could produce the minimal genome than mindless natural processes. Again, if one wishes to explain the origin of the simplest life form by natural selection, a fitness function will be required that is capable of generating 267,000 bits of functional information, well into the area that requires intelligent design."* [Kal08p11] Note that even if factors of 10^7 more trials and 1000 more functional results, and the minimum genes were halved, ID would still be $10^{40,130}$ more probable than mindless natural processes as the cause of the simplest life. Also keep in mind that since life would not have yet begun, any natural section would have to be operative on non-life. Also, recall that other factors, such as thermodynamics, enzymes, etc. are not considered.

Appendix D: Scientific Falsification and Specific Null Hypotheses

"The criterion of the scientific status of a theory is its falsifiability, or refutability, or testability" [Pop63]. See page 3 for Universal Plausibility Metric [Abe09U].

Shannon Channel Capacity Theorem [Sha48] would require falsification before giving consideration as science to any scenario proposing an alphabet with lower symbolic complexity than the current codon alphabet (see Appendix C).

Testable Null Hypotheses: A single incident would falsify any of these hypotheses.
[Abe05] Testable Null hypotheses on Functional Sequence Complexity:
#1 Stochastic ensembles of physical units cannot program algorithmic/cybernetic function.
#2 Dynamically-ordered sequences of individual physical units (physicality patterned by natural law causation) cannot program algorithmic/cybernetic function.
#3 Statistically weighted means (e.g. increased availability of certain units in the polymerization environment) giving rise to patterned (compressible) sequences of units cannot program algorithmic/cybernetic function.
#4 Computationally successful configurable switches cannot be set by chance, necessity, or any combination of the two, even over large periods of time.

[Abe06] Testable Null Hypotheses About Cybernetic Organization
#1 Self-ordering phenomena cannot generate cybernetic organization.
#2 Randomness cannot generate cybernetic organization.

[Abe09P] Testable Null Hypotheses About Prescriptive Information (PI)
#1 PI cannot be generated from/by the chance and necessity of inanimate physicodynamics.
#2 PI cannot be generated independent of formal choice contingency.
#3 Formal algorithmic optimization, and the conceptual organization that results, cannot be generated independent of PI.

[Abe09C] Testable Null Hypothesis About the Cybernetic Cut
Physicodynamics cannot spontaneously traverse The Cybernetic Cut: physicodynamics alone cannot organize itself into formally functional systems requiring algorithmic optimization, computational halting, and circuit integration.

References (alphabetized by last name of primary author, if any)

AAAS-4/11/05, "Harvard's Nima Arkani-Hamed Ponders New Universes, Different Dimensions," www.aaas.org/news/releases/2005/0511string.shtml

Abbot (Larry) , "The Mystery of the Cosmological Constant," Scient. American: 3 (1), 1991.

Abel (David) & Jack Trevors, "Three Subsets of Sequence Complexity and Their Relevance to Biopolymeric Information," Theoretical Biology and Medical Modelling, 8/11/05, 2:29.

Abel (David) & Jack Trevors, "Self-Organization vs Self-Ordering Events in Life-Origin Models," Physics of Life Reviews (3), 2006, p211-228.

Abel (David), "Complexity, Self-organization, and Emergence at the Edge of Chaos in Life-Origin Models," Journal of the Washington Academy of Sciences 93 (4), 2007, p1-20.

Abel (David), "The 'Cybernetic Cut': Progressing from Description to Prescription in Systems Theory," The Open Cybernetics and Systemics Journal (2), 2008, p252-262.

Abel (David), "The GS (genetic selection) Principle," Frontiers in Bioscience (14), 1/1/09G, p2959-2969.

Abel (David), "The Biosemiosis of Prescriptive Information," Semiotica, 1/4/09P, p1-19.

Abel (David), "The Capabilities of Chaos and Complexity," Int. J. Mol. Sci. (10), 2009C, p247-291.

Abel (David), "The Universal Plausibility Metric (UPM) & Principle (UPP)," Theoretical Biology and Medical Modelling, 12/3/09U, 6:27.

Abel (David), "Constraints vs Controls," The Open Cybernetics & Systemics Journal (4), 2010, p14-27.

Abelson (P.), "Chemical Events on the Primitive Earth," PNAS USA: 55, 1966, p1365-1372.

Adami (C.), "Ab Initio Modeling of Ecosystems with Artificial Life," Natural Resource Modeling: 15, 2002, p133-146.

Adleman (Leonard), "Molecular Computation Of Solutions To Combinatorial Problems," Science: 266, 11/11/94, p1021–1024.

Alberts (Bruce), Molecular Biology of the Cell, 1994, p533.

Alfven (Hannes) , "Cosmology in the Plasma Universe - an Introductory Exposition," IEEE Transactions on Plasma Science: 18, 2/90, p5-10.

Ans-Web, http://www.answers.com/topic/overlapping-genes

Appeals Court, 7th Circuit, Kaufman, James v. McCaughtry & Gary, 8/20/05.

Appeals Court, Comer v. TEA, 5th Circuit, 7/2/10.

Arnold (J.) & W. Libby, "Age Determinations by Radiocarbon Content: Checks with Samples of Known Age," Science: 110 (2869), 1949, p678–680.

Arrhenius (S.), Worlds in the Making, 1908.

Asimov (Isaac), "Can Decreasing Entropy Exist in the Universe?," Sci. Digest, 5/73, p76-77.

Astrobiology Magazine, "Inevitability Beyond Billions," 7/03.
Astrophysics Spectator: 2.33, 10/5/05.
Avida Manual, www.krl.caltech.edu/~charles/avida/manual/intro.html
Avida Website, www.krl.caltech.edu/avida/home/software.html
Axe (Doug), "Estimating the Prevalence of Protein Sequences Adopting Functional Enzyme Folds," J Mol Biol. 8/27/04, p1295-1315.
Axe (Doug), "The Case Against a Darwinian Origin of Protein Folds," BIO-Complexity, 2010(1), p1-12.
Ayala (Francisco), "Darwin's Revolution" in Creative Evolution?!, 1994, p3-5.
Babaoglu (O.), M. Jelasity, G. Canright, T. Urnes, A. Deutsch, N. Ganguly, G. Di Caro, F. Ducatelle, L. Gambardella, R. Montemanni, & A. Montresor, "Design Patterns from Biology for Distributed Computing," ACM Trans on TAAS, 5/9/06, p26-66.
Bada (Jeffrey), "How Life Began on Earth: a Status Report," Earth and Planetary Science Letters: 226, 9/30/04, p1-15.
Bandyopadhyay (A.), R. Pati, S. Sahu, F. Peper & D. Fujita, "Massively Parallel Computing on an Organic Molecular Layer," Nature Physics 6, 4/25/10, p369-375.
Barash (Y.), J. Calarco, W. Gao, Q. Pan, X. Wang, O. Shai, B. Blencowe, & B. Frey, "Deciphering the Splicing Code," Nature, 5/6/10, p53-9.
Barbieri (Marcello), "Biosemiotics: a New Understanding of Life," Naturwissenschaften (95), 2/19/08B, p577-599.
Barbieri (Marcello), "Life is Semiosis – The Biosemiotic View of Nature," Cosmos and History: The Journal of Natural and Social Philosophy (4), 2008S, p29-51.
Barnes (R. K.), P. Calow & P. W. Olive, The Invertebrates: A New Synthesis, 2001, p9–10.
Barricelli (Nils Aall), Symbiogenetic Evolution Processes Realized by Artificial Methods, Methodos, 1957, p143–182.
Barrow (John) & Joseph Silk, "The Structure of the Early Universe," Scientific American, 4/80, p118-128.
Barry (Patrick), "Life from Scratch," Science News Online: 173 (2), 1/12/08L, p27.
Battail (Gerard), "Genetics as a Communication Process Involving Error-Correcting Codes," in Biosemiotics: Information, Codes and Signs in Living Systems, 2007, p105.
Behe (Michael), Darwin's Black Box: the Biochemical Challenge to Evolution, 1996.
Behe (Michael), "Irreducible Complexity and the Evolutionary Literature: Response to Critics," 7/31/00, www.arn.org/docs/behe/mb_evolutionaryliterature.htm
Benenson (Y.), B. Gil, U. Ben-Dor, R. Adar, & E. Shapiro, "An Autonomous Molecular Computer for Logical Control of Gene Expression," Nature: 429, 4/28/04, p423–429.
Benner (S.), H. Kim, M. Kim, & A. Ricardo, "Planetary Organic Chemistry and the Origins of Biomolecules," Cold Spring Harb Perspect Biol, 6/1/10.

Bennett (C. H.), "Logical Reversibility of Computation," IBM Jour of Research & Development: 17, 1973, p525-532

Bergman (Jerry), Vestigial Organs Are Fully Functional, 1990.

Bergman (Jerry), Slaughter of the Dissidents: The Shocking Truth about Killing the Careers of Darwin Doubters, 2008.

Bernstein (M.), J. Dworkin, S. Sandford, G. Cooper, & L.Allamandola, "Racemic Amino Acids from the Ultraviolet Photolysis of Interstellar Ice Analogues," Nature: 416, 3/28/02, p401-403.

Bernstein (Max), "Prebiotic Materials From on and off the Early Earth," Phil. Trans. R. Soc. B:361, 2006P, p1689–1702.

Berry (Richard), in "Bacterial Motors Could Inspire Nanotechnology," 2/20/06B, www.physorg.com/news11029.html

Biomass, Wikipedia, http://en.wikipedia.org/wiki/Biomass_(ecology)

Boneh (D.), C. Dunworth, R. Lipton, & J. Sgall, "On the Computational Power of DNA," DAMATH 11, 1996, www.dna.caltech.edu/courses/cs191/paperscs191/bonehetal.pdf

Borek (Ernest), The Sculpture of Life, Columbia Univ Press, 1973, p5.

Borel (Emil), Probability and Certainty, 1950.

Borman (Stu), "Protein Factory Reveals Its Secrets," C&E News: 85(8), 2/19/2007, p13-16.

Bostrom (Nick), Anthropic Bias: Observation Selection Effects in Science and Philosophy, 2002.

Bowman (Lee), "Open Inquiry: the New Science Standard," Uncommon Descent, 10/20/07, www.uncommondescent.com/education/open-inquiry-the-new-science-standard

Bracht (John), "The Bacterial Flagellum: A Response to Ursula Goodenough," 2003, www.iscid.org /papers/Bracht_GoodenoughResponse_021203.pdf

Brady (Ronald), "On the Independence of Systematics," Cladistics: 1, 1985, p113-126.

Brumfiel (Geoff), "Outrageous Fortune," Nature, 1/5/06, p10-12.

Bruni (L.), "Cellular Semiotics and Signal Transduction," in Introduction to Biosemiotics, 2007, p365-407.

Buchanan (Mark), "Horizontal and Vertical: the Evolution of Evolution," New Scientist, 1/26/10, #2744.

Cairns-Smith (A.G.), Seven Clues to the Origin of Life: A Scientific Detective Story, 1993, p44-45.

Calderone (Melissa), "Do You Use More Energy When You're Thinking Really Hard?," www.popsci.com/scitech/article/2006-07/mental-workout

Calvert (John), "Kitzmiller's Error: Defining Religion Exclusively rather than Inclusively," Liberty University Law Review: 3(2), Spr 2009, p213-328.

Cannarozzi (G.), N. Schraudolph, M. Faty, P. von Rohr, M. Friberg, A. Roth, P. Gonnet, G. Gonnet, & Y. Barral, "Role for Codon Order in Translation Dynamics," Cell: 141, 4/16/10, p355-367.

Chaitin (Gregory), "Toward a Mathematical Definition of Life," in The Maximum Entropy Formalism, 1979.

Chaitin (Gregory), "Speculations on Biology, Information and Complexity," EATCS Bulletin: 91, 2/07, p231-237.

Chinneck (John), "Analyzing Infeasible Optimization Models," at CORS/INFORMS, 5/04.

Clarkson (Euan) & Riccardo Levi-Setti, "Trilobite Eyes and the Optics of Des Cartes and Huygens," Nature: 254, 4/24/75, p663-667.

Cohen (S.), A. Chang, H. Boyer, & R. Helling, "Construction of Biologically Functional Bacterial Plasmids In Vitro," PNAS USA: 70 (11), 11/73, p3240-3244.

Cooper (G.), N. Kimmich, W. Belisle, J. Sarinana, K. Brabham, & L. Garrel, "Carbonaceous Meteorites as a Source of Sugar-related Organic Compounds for the Early Earth," NASA Tech Report 20040088530, 2001.

Corning (Peter) & Stephen Kline, "Thermodynamics, Information and Life Revisited, Part I: to Be or Entropy," Systems Research, 4/7/00, p273-295.

Corning (Peter), Holistic Darwinism, 2005, p330.

CosmicRays.org, "Solar Neutrinos - Calculation of the solar neutrino flux on Earth," www.cosmicrays.org/muon-solar-neutrinos.php

Costanzo (G.), R. Saladino, C. Crestini, F. Ciciriello, & E. DiMauro, "Nucleoside Phosphorylation by Phosphate Minerals," J. Biol. Chem: 282 (23), 6/8/07, 16729-16735.

Crews (Frederick),"Saving Us from Darwin," NY Review of Books: 48 (15), 10/4/01.

Crick (Francis), "The Origin of the Genetic Code," J Mol Biol: 38, 1968, p367–379.

Crick (Francis), "Directed Panspermia," Icarus: 19, 1973, p341-346.

Crocker (Caroline), Free To Think, 2010.

Darnell (J.), H. Lodish, & D. Baltimore, Molecular Cell Biology, 1986.

Darwin (Charles), Origin of Species, Paperback (of 1859), 1998, p154.

Das (Saswato) , "Physicists Make Artificial Black Hole Using Optical Fiber," IEEE Spectrum, 3/08, http://spectrum.ieee.org/mar08

Davenport (John) "Possible Progenitor of DNA Re-Created," Science Now, 11/16/00, p1.

Davies (Paul), Superforce: The Search for a Grand Unified Theory, 1984, p235-236.

Davies (Paul), The Cosmic Blueprint: New Discoveries in Nature's Creative Ability To Order the Universe, 1988, p203.

Davies (Paul), Wiki-Quote, http://en.wikiquote.org/wiki/Darwinism

Davis (Jimmy) & Harry Poe, Designer Universe, 2002.

Dawkins (Richard), The Selfish Gene, 1976.

Dawkins (Richard),"The Necessity of Darwinism," New Scientist: 94, 4/15/82, p130.

Dawkins (Richard), Scientific American, 6/88.

Dawkins (Richard), "Book Review," The New York Times, 4/9/89, section 7, p3.

Dawkins (Richard), River Out of Eden,1995, p83.
Dawkins (Richard), The Blind Watchmaker, 1996B.
Dawkins (Richard), Climbing Mount Improbable, 1996C.
Dawkins (R.), A Devil's Chaplain: Reflections on Hope, Lies, Science, and Love, 2001, p79.
Dawkins (Richard), A Devil's Chaplain: Reflections on Hope, Lies, Science, and Love (paperback), 2004, p99.
Dawkins (Richard), "The Information Challenge," 1998 & 2008I, www.skeptics.com.au/articles/dawkins.htm
Dawkins (Richard), "Lying for Jesus?,"3/23/08L, http: //richarddawkins.net /article,2394,Lying-for-Jesus,Richard-Dawkins.
De Duve (C.),"The Beginning of Life on Earth," American Scientist, 1995.
Dembski (William) The Design Inference, 1998.
Dembski (William), Intelligent Design, 1999.
Dembski (William) "Reinstating Design Within Science," in, Unapologetic Apologetics, 2001.
Dembski (William), "Irreducible Complexity Revisited," 2004, www.designinference.com/documents/2004.01.Irred_Compl_Revisited.pdf
Denton (Michael), Evolution: A Theory in Crisis, 1986.
Denton (Michael), Nature's Destiny: How the Laws of Biology Reveal Purpose in the Universe, 1998.
DeRosier (D.), "Spinning Tails," Curr Opin Struct Biol., 4/5/95, p187-93.
Deutsch (David), Interviewed on The Science Show: The Anthropic Universe, 2/18/06.
DeWolf (David), John West, & Casey Luskin, "Intelligent Design Will Survive Kitzmiller v. Dover," 68 Montana Law Review 7, Spring, 2007, p30.
Diaconis (Persi), Susan Holmes, & Richard Montgomery, "Dynamical Bias in the Coin Toss," SIAM Review: 49 (2), 4/07, p211-235.
Discovery staff, "The Truth About the Dover Intelligent Design Trial," 11/15/07, www.discovery.org/a/2879
Dissent-web, "A Scientific Dissent From Darwinism," www.dissentfromdarwin.org/.
D'Onofrio (David) & Gary An, "A Comparative Approach for the Investigation of Biological Information Processing: an Examination of the Structure and Function of Computer Hard Drives and DNA," Theoretical Biology and Medical Modelling, 2010, 7:3.
Dose (Klaus), "The Origin of Life: More Questions Than Answers," Interdisciplinary Science Reviews: 13 (4), 1988, p348.
Durston (Kirk), David Chiu, David Abel, & Jack Trevors, "Measuring the Functional Sequence Complexities of Proteins," Theo. Bio. and Medical Modelling, 12/6/07, 14 pgs.
Dyson (Lisa), Matthew Kleban, & Leonard Susskind, "Disturbing Implications of a Cosmological Constant," JHEP, 2/10/02, p11.

Easterbrook (Gregg), "Science and God: A Warming Trend?," Science, 8/97, p890-893.
Easterbrook (Gregg), "Where did life come from?," Wired Magazine, 2/07, p108.
Eddington (Arthur), Space, Time and Gravitation: An Outline of the General Relativity Theory, 1920.
Eden (Murray), "Inadequacies of Neo-Darwinian Evolution as a Scientific Theory," Mathematical Challenges to the Neo-Darwinian Interpretation of Evolution, Wistar Institute, 1966.
Edinburgh University, "Insight into Cells Could Lead to New Approach to Medicines," Science Daily, 6/22/10.
Eiben (A.) & J. Smith, Introduction to Evolutionary Computing, 2003.
Einstein (Albert), Ideas and Opinions, 1954, p255.
Ehrlich (P.) & L. Birch, "Evolutionary History and Population Biology," Nature, 4/22/67, p352.
Eldredge (Niles) & Stephen Jay Gould, "Punctuated Equilibria: an Alternative to Phyletic Gradualism," in T.J.M. Schopf, ed., Models in Paleobiology,1972 p82-115.
Eldredge (Niles), "A Trilobite Panorama in Eastern North America," Fossils Magazine: 1, 1976, p58-67.
Eldredge (Niles), Harper's Magazine, 2/85, p60.
Emmeche (Claus), "Defining Life, Explaining Emergence," 1997, www.nbi.dk/~emmeche/cePubl/97e.defLife.v3f.html
Entrez Database, www.ncbi.nlm.nih.gov/sites/entrez?db=protein
ERPANET/CODATA Workshop, "The Selection, Appraisal and Retention of Digital Scientific Data," Biblioteca Nacional, Lisbon, 12/15-17/03.
Evolution-site, http://evolution.berkeley.edu/evosite/evo101/IIntro.shtml.
Ewert (W.), G. Montañez, W. Dembski, R. Marks II, " Efficient Per Query Information Extraction from a Hamming Oracle," Proceedings of the the 42nd Meeting of the Southeastern Symposium on System Theory, IEEE, University of Texas at Tyler, 3/7-9/10, p290-297.
Expelled, The Movie, 2008.
Faria (M.), "RNA as Code Makers: A Biosemiotic view of RNAi and Cell Immunity," in Introduction to Biosemiotics, 2007, p 347-364.
Ferris (J.), A. Hill Jr, R. Liu, & L. Orgel, "Synthesis of Long Prebiotic Oligomers on Mineral Surfaces," Nature: 381, 1996, p59-61.
Flynn (M.), "Some Computer Organizations and Their Effectiveness," IEEE Trans. Comput: C-21, 1972, p948 .
FM-tribolites, www.fossilmuseum.net/Evolution/TrilobiteArmsRace.htm
Fodor (J.) & M. Piattelli-Palmarini, "Survival of the Fittest Theory: Darwinism's Limits," New Scientist, 2/3/10S.
Fodor (J.) & M. Piattelli-Palmarini, What Darwin Got Wrong, 2010W.
Forster (Anthony) & George Church, "Towards Synthesis of a Minimal Cell,"

Molecular Systems Biology: Article 45, 8/22/06.
Fraser (C.M.), et al., "The Minimal Gene Complement of Mycoplasma Genitalium," Science: 270 (5235), 1995, p397-403.
Freeland (Stephen), Robin Knight, Laura Landweber, & Laurence Hurst, "Early Fixation of an Optimal Genetic Code," Molecular Biology and Evolution: 17, 2000, p511-518.
Frey (J.), quoted in "Researchers Crack 'Splicing Code,' Solve a Mystery Underlying Biological Complexity," 5/5/10, www.physorg.com/news192282850.html
F-theory, http://en.wikipedia.org/wiki/F-theory
Gal (J.), G.Horvath, E. Clarkson, & O. Haiman, "Image formation by Bifocal Lenses in a Trilobite Eye?," Vision Research: 40, 2000, p843–853.
Gange (Robert), Origins and Destiny, 1986, p77.
Gates (Bill), The Road Ahead, [1995], Revised, 1996, p.228.
Gee (Henry), In Search of Deep Time, 1999, p113-117.
Gefter (Amanda), "Is String Theory in Trouble?," New Scientist, 12/17/05.
Gehring (W. J.), "New Perspectives on Eye Development and the Evolution of Eyes and Photoreceptors," Journal of Heredity: 96 (3), 2005, p171-184.
Geoclassics, "Trilobites," www.geoclassics.com/trilobites.htm.
Gibbs (W.), "The Unseen Genome: Gems Among the Junk," Sci. American, 11/03, p46-53.
Gibson (D.), J. Venter, & 21 others, "Creation of a Bacterial Cell Controlled by a Chemically Synthesized Genome," Science Express, 5/20/10, p1-12.
Gilder (George), "Evolution and Me," National Review, 7/17/06.
Gimona (M.), "Protein Linguistics and the Modular Code of the Cytoskeleton," in The Codes of Life: The Rules of Macroevolution, 2008, p189-206.
Gingeras (Thomas), www.desdeelexilio.com/2010/06/28/epigenetica-entrevista-a-thomas-gingeras/
Gitt (Werner), In the Beginning was Information, 1997.
Glass (J.), J. Venter, & 7 others, "Essential Genes of a Minimal Bacterium," PNAS: 103, 2006, p425-430.
GMIS (US Energy Department Genome Management Information System), http://genomics.energy.gov.
Gon (S. M.), "The Trilobite Eye," 10/1/07, www.trilobites.info/eyes.htm
Gorman (Jessica), "Cosmic Chemistry Gets Creative," Sci. News: 159 (20), 5/19/01, p318.
Grassé (Pierre-P), Evolution of Living Organisms, 1977.
Gregory (R. L.), Eye and Brain: The Psychology of Seeing, second edition, 1972, p25.
Gribbin (John), Are We Living in a Designer Universe," UK Telegraph, 8/31/10.
Grünwald (David) and Robert Singer, "In Vivo Imaging of Labelled Endogenous (-actin mRNA During Nucleocytoplasmic Transport," Nature, 2010, DOI: 10.1038/nature09438.

Haeckel (Ernst), The Evolution of Man, Translation (original 1911) 2004, p49.
Haldane J.B.S.), "The Cost of Natural Selection," J. Genet.: 55, 1957, p511-524.
Hall (Brian), "Baupläne, Phylotypic Stages, and Constraint: Why There Are So Few Types of Animal," Evolutionary Biology: 29, 1996,
Hamming (R. W.), "Error Detecting and Correcting Codes," Bell Sys Tech J, 1950, p147-160.
Hanczyc (M.), S. Mansy, & J. Szostak, "Mineral Surface Directed Membrane Assembly," Orig Life Evol Biosph: 37 (1), 2/07, p67-82.
Hannon, (Gregory), quoted in "Messenger RNAs Are Regulated in Far More Ways than Previously Appreciated," Science Daily, 6/25/10.
Hao (B.), W Gong, T Ferguson, C James, J Krzycki, & M Chan, "A new UAG-encoded Residue in the Structure of a Methanogen Methyltransferase," Science, 5/24/02, p1462-1466.
Harold (F.), The Way of th e Cell: Molecules, Organisms and the Order of Life, 2001, p205.
Hawking (Stephen), A Brief History of Time, 1988.p205.
Hawking (Stephen) & Leonard Mlodinow, The Grand Design, 2010.
Hayden (Erika), "Human Genome at Ten: Life is Complicated," Nature 464, 3/31/10, p664-667.
Hazen (Robert), "Genesis: Rocks, Minerals, and the Geochemical Origin of Life," Elements: 1, 6/05, p135–137.
Hazen (Robert), Patrick Griffin, James Carothers, & Jack Szostak, "Functional Information and the Emergence of Biocomplexity," PNAS: 104-1, 5/15/07, p8574-8581.
Hazen (Robert) & Dimitri Sverjensky, "Mineral Surfaces, Geochemical Complexities, and the Origins of Life," Cold Spring Harb Perspect Biol, 4/14/10, 2:a002162.
Hintz (M.), C. Bartholmes, P. Nutt, J. Ziermann, S. Hameister, B. Neuffer, & G. Theissen, "Catching a 'Hopeful Monster': Shepherd's Purse (Capsella Bursa-pastoris) as a Model System to Study the Evolution of Flower Development," Journal of Experimental Botany 57(13), 2006 p3531-3542.
Hoffmeyer (J.) & C. Emmeche, "Code-Duality and the Semiotics of Nature," J. Biosemiotics (1), 2005, p37-64.
Horgan (John), "In The Beginning...," Scientific American: 264 (2), 2/91, p100-109.
Horgan (John), "The Consciousness Conundrum," IEEE Spectrum Online, 6/08.
Hoyle (Fred), "The Universe: Past and Present Reflections," Engineering and Science, 11/81U, p8-12.
Hoyle (Fred) & Chandra Wickramasinghe, Evolution from Space, 1981E.
Hoyle (Fred), "Evolution from Space,"Omni Lecture at the Royal Institution, London, 1/12/82.
Hoyle (Fred), The Intelligent Universe, 1983, p16-17.
Hoyle (F.), G. Burbidge, & J. Narlikar, Astrophysical Journal: 1-410 (2), 06/93, p437-457.
Hoyle (F.), G Burbidge & J Narlikar, "The Basic Theory Underlying the Quasi-Steady State Cosmological Model," Proc. R. Soc. A: 448, 1995, p191.

Hoyle (Fred), Mathematics of Evolution, 1999, book cover.
Hubble (Edwin), "A Relation Between Distance and Radial Velocity Among Extra-galactic Nebulae," Proceedings of the National Academy of Sciences: 15, 1929, p168–173.
Hunter (Cornelius), Darwin's Proof, 2003, p60.
ID-Web, "What is intelligent design?," www.intelligentdesign.org/whatisid.php.
Itaya (Mitsuhiro), "An Estimation of the Minimal Genome Size Required for Life," FEBS Letters: 362, 1995, p257–60.
Jenuwein (T.) & C. Allis, "Translating the Histone Code," Science, 8/10/01, p1074-1080.
Jimenez-Montano (Miguel), "Applications of Hyper Genetic Code to Bioinformatics," J. Biol. Sys.: 12, 2004, p5-20.
Johnson (D.), D. Lilja & J. Riedl, "A Circulating Active Barrier Synchronization Mechanism," International Conference on Parallel Processing:I, 8/95, p202-209.
Johnson (D.), D. Lilja, J. Riedl,& J. Anderson, "Low-Cost, High-Performance Barrier Synchronization on Networks of Workstations," J Par & Distrib Comp, 2/97B, p131-137.
Johnson (Donald), Exploring Fine-Grained Process Interaction in Multiprocessor Systems, University of Minnesota Thesis, 1997T.
Johnson (Donald), "Data and Information: Effect of Bioinformatics on Traditional Biology," Int Conf on Bioinformatics (Poster), 12/04.
Johnson (D.), D. Lilja & J. Riedl, "Circulating Shared-Registers for Multiprocessor Systems," Jour. Of Systems Arch., 6/3/05, p152-168.
Johnson (Phillip), "The Church of Darwin," The Wall Street Journal, 8/16/99.
Joyce (G.), G. Visser, C. van Boeckel, J. van Boom, L. Orgel, & J. van Westrenen, "Chiral Selection in Poly(C)-directed Synthesis of Oligo(G)," Nature: 310, 1984, p602-604.
Joyce (Gerald), "Nucleic Acid Enzymes: Playing with a Fuller Deck," PNAS: 95 (11), 5/26/98, p5845–5847.
Joyce (Gerald) & Leslie Orgel, "Prospects for Understanding the Origin of the RNA World," in The RNAWorld, 2nd ed.,1999, p49–77.
Kalinsky (K. D.), "Intelligent Design: Required by Biological Life?," 2/19/08, http://www.newscholars.com/papers/ID%20Web%20Article.pdf.
Kauffman (Louis), CYBCON discusstion group 18, 9/20/07, p15.
Keith (Arthur), Evolution and Ethics, 1947, p230.
Keplinger (B.L.), A.L. Rabetoy, & D.R. Cavener, "A Somatic Reproductive Organ Enhancer Complex Activates Expression in Both the Developing and the Mature Drosophila Reproductive Tract," Developmental Biology: 180, 1996, p311-323.
Kim (Y.), M, Coppey, R. Grossman, L. Ajuria, G. Jiménez, Z. Paroush, S. Shvartsman, "MAPK Substrate Competition Integrates Patterning Signals in the Drosophila Embryo," Current Biology: 20, 3/9/10, p1-6.

Kinoshita (S.), S. Kageyama, K. Iba, Y. Yamada, & H. Okada,, "Utilization of a Cyclic Dimer and Linear Oligomers of e-aminocaproic Acid by Achromobacter Guttatus," Agricultural & Biological Chemistry (39), 6/75, p1219-23.

Klug (Stefanie), Alexander Hüttenhofer, Matthias Kromayer, & Michael Famulok, "In Vitro and in Vivo Characterization of Novel mRNA Motifs that Bind Special Elongation Factor SelB," Proc.Natl. Acad. Sci. USA: 94, 6/97, p6676-6681.

Kohler (J.), S. Schafer-Preuss, & D. Buttgereit, "Related Enhancers in the Intron of the Beta1 Tubulin Gene of Drosophila Melanogaster are Essential for Maternal and CNS-specific Expression During Embryogenesis," Nucleic Acids Research: 24, 1996, p2543-2550.

Kondo (T.), S. Plaza, J. Zanet, E. Benrabah, P. Valenti, Y. Hashimoto, S. Kobayashi, F. Payre, Y. Kageyama, "Small Peptides Switch the Transcriptional Activity of Shavenbaby During Drosophila Embryogenesis," Science 329, 7/16/10, p336-339.

Koonin (Eugene) & Artem Novozhilov, "Origin and Evolution of the Genetic Code: The Universal Enigma," arXiv:0807.4749, 7/08.

Koonin (Eugene), "Darwinian Evolution in the Light of Genomics," Nucleic Acids Research 37(4), 2/12/09, p1011–1034.

Koons (Robert), "Are Probabilities Indispensable to the Design Inference?," 5/22/01, www.utexas.edu/cola/depts/philosophy/faculty/koons/ontocomplex.pdf

Kormendy (J.), D. Richstone, "Inward Bound---The Search For Supermassive Black Holes In Galactic Nuclei," Annual Reviews of Astronomy and Astrophysics: 33, 1995, p581-624.

Kozmik (Z.), J. Ruzickova, K. Jonasova, Y. Matsumoto, P. Vopalensky, I. Kozmikova, H. Strnad, S. Kawamura, J. Piatigorsky, V. Paces, & C. Vlcek, "Assembly of the Cnidarian Camera-Type Eye from Vertebrate-Like Components," PNAS, 7/1/08, p8989–8993.

Kruger (K.), P. Grabowski, A. Zaug, J. Sands, D. Gottschling, & T. Cech, "Self-Splicing RNA: Autoexcision and Autocyclization of the Ribosomal RNA Intervening Sequence of Tetrahymena," Cell 31(1), 11/82, p147-157.

Kull (Kalevi), "A Brief History of Biosemiotics," in Biosemiotics: Information, Codes and Signs in Living Systems, 2007, p2.

Lad (Chetan), Nicholas Williams, & Richard Wolfenden, "The Rate of Yydrolysis of Phosphomonoester Dianions and the Exceptional Catalytic Proficiencies of Protein and Inositol Phosphatases," PNAS: 100 (10), 5/13/03, p5607-5610.

Lane (N.), J. Allen, & W. Martin, "How Did LUCA Make a Living? Chemiosmosis in the Origin of Life," Bioessays, 1/27/10, 10.1002/bies.200900131.

Lenski (Richard), Charles Ofria, Robert T. Pennock, & Christoph Adami, "The Evolutionary Origin of Complex Features," Nature: 423, 5/8/03, p139-44.

Le Page (M.), "Genome at 10: A Dizzying Journey into Complexity," New Scientist, 6/16/10.

Lester (Lane), James C. Hefley, Human Cloning: Playing God or Scientific Progress?, 1998.

Levi-Setti (R.), Trilobites: A Photographic Atlas (2nd edition), 1993.
Lewin (Roger), "Evolutionary Theory under Fire," Science: 210, 1980, p883.
Lewis (Ricki), Bruce Parker, Douglas Gaffin, Marielle Hoefnagels, Life, 2006, Sec 13.5.
Lewontin (Richard), "Billions and Billions of Demons," in The NY Review of Books, 1/9/97.
Lieberman (Judy), "Master of the Cell," The Scientist, 4/1/10, p42.
Lloyd (Seth), "Computational capacity of the Universe," Phys. Rev. Lett.: 88, 5/24/02, 4 pgs.
Loeb (Walther), "The Effect of Silent Discharge on the Reactions of Formamide A Contribution to the Question of Nitrogen Assimilation," Rep of German Chem Soc, 1913, p684-697.
Lovgren (Stefan), "Computer Made from DNA and Enzymes," National Geographic, 2/24/03, news.nationalgeographic.com/news /2003/02/0224_030224_DNAcomputer.html
Ludwig (Mark), Computer Viruses, Artificial Life and Evolution, 1993.
Luisi (P.), "The Problem of Macromolecular Sequences: the Forgotten Stumbling Block," Origins of Life and Evolution of the Biosphere 37, 4–5/2007, p363–365.
Luskin (Casey), "The Facts about Intelligent Design," 2008, www.ideacenter.org/contentmgr/showdetails.php/id/1452
Luskin (Casey), "Does Challenging Darwin Create Constitutional Jeopardy? A Comprehensive Survey of Case Law Regarding the Teaching of Biological Origins," Hamline University Law Review: 32(1), 2009, p1-64 (link: www.discovery.org/a/11291).
MacDónaill (Dónall), "Digital Parity and the Composition of the Nucleotide Alphabet," IEEE Engineering in Medicine and Biology, 1-2/06, p54-61.
Macnab (Robert), CRC Crit. Rev. Biochem: 5, 12/78, p333.
Maddox (John), What Remains to Be Discovered: Mapping the Secrets of the Universe, the Origins of Life, and the Future of the Human Race, 1998, p252.
Madrigal (Alexis), "Wired Science Reveals Secret Codes in Craig Venter's Artificial Genome," 1/28/08, blog.wired.com/wiredscience/2008/01/venter-institut.html.
Makalowski (Wojciech), "Not Junk After All.," Science, 5/23/03.
Mansy (Sheref), "Membrane Transport in Primitive Cells," Cold Spring Harb Perspect Biol, 4/21/10, doi: 10.1101
Margulis (Lynn) & Dorion Sagan, Acquiring Genomes: A Theory of the Origins of the Species, 2003, p29.
Mattick (John), "Challenging the Dogma: The Hidden Layer of Non-Protein-Coding RNAs in Complex Organisms," BioEssays: 25, 10/03, p930–939.
May (E.), M. Vouk, D. Bitzer, & D, Rosnick, "An Error-Correcting Code Framework for Genetic Sequence Analysis," J. Frank Inst.: 341, 1-3/04, p89-109.
Mazur (Suzan), The Altenberg 16: An Exposé of the Evolution Industry, 2010.
McCormick (T.) & R.Fortey, "Independent Testing of a Paleobiological Hypothesis: the Optical Design of Two Ordovician Pelagic Trilobites Reveals Their Relative Paleobathymetry,"

Paleobiology: 24 (2), 1998, p235–253.
McDonald (John), "The Molecular Basis of Adaptation: A Critical Review of Relevant Ideas and Observations," Annual Review of Ecology and Systematics: 14, 1983, p77–102.
McIntosh (Andy), "Entropy, Free Energy and Information in Living Systems," International Journal of Design & Nature and Ecodynamics 4 (4), 2009, p351-385.
Meyer (Stephen), "The Explanatory Power of Design: DNA and the Origin of Information," Mere Creation Conf., 1998, p114.
Meyer (Stephen), "DNA and Other Designs," First Things: 102, 4/1/00, p30-38.
Meyer (Stephen), Marcus Ross, Paul Nelson, & Paul Chien, "The Cambrian Explosion: Biology's Big Bang," Darwinism, Design and Public Education, 2003.
Meyer (Stephen), "The Origin of Biological Information and the Higher Taxonomic Categories," in Proc Biol Soc Wash: 117 (2), 2004, p213-239.
Meyer (Stephen), Signature in the Cell, 2009.
Michael (Eli), "How Physically Plausible is the Cosmological Constant?," from the University of Colorado, Boulder, 1999, super.colorado.edu/~michaele/Lambda/phys.html
Miller (S. L.), "Production of Amino Acids Under Possible Primitive Earth Conditions," Science: 117, 1953, p528.
Miller (S. L.) & L. E. Orgel, The Origins of Life on the Earth, 1974, p33.
Mims (Forrest), Rejected Letter, Science, www.forrestmims.org/publications.html,12/94.
Minnich (Scott) & Stephen C. Meyer, "Genetic Analysis of Coordinate Flagellar and Type III Regulatory Circuits in Pathogenic Bacteria," Proc 2nd Int Conf. on Design & Nature, 9/04.
Minnich (Scott), in Unlocking the Mystery of Life (video).
Moore's Law Webpage, www.intel.com/technology/mooreslaw/, 1965 graph.
Morowitz (Harold), Energy Flow in Biology, 1979, p99.
Morris (Conway), "Evolution: Bringing Molecules into the Fold," Cell: 100, 1/7/00, p1-11.
Morris (Conway), "Eyes to See, Brains to Think: the Inevitable Evolution of Intelligence," 2/20/07, www.hss.ed.ac.uk/Admin/Gifford/documents /SimonConwayMorristitleandabstracts3.pdf
Morris (M.), K. Thorne, & U. Yurtsever, "Wormholes, Time Machines, and the Weak Energy Condition," Physical Review: 61 (13), 9/88, p1446-1449.
Murray (Daniel) & Scott W. Teare, "Probability of a Tossed Coin Landing on Edge," Phys. Rev. E 48,10/93, p2547-2552.
Murtas (Giovanni), "Construction of a Semi-Synthetic Minimal Cell: A Model for Early Living Cells," Orig Life Evol Biosph: 37, 10/07C, p419-422.
Murtas (Giovanni), Yutetsu Kuruma, Paolo Bianchini, Alberto Diaspro, Pier Luigi Luisi, "Protein Synthesis in Liposomes with a Minimal Set of Enzymes," Biochem Biophys Res Commun, 11/9/07P, p12-17.

Mushegian (Arcady) & Eugene Koonin, "A Minimal Gene Set for Cellular Life Derived by Comparison of Complete Bacterial Genomes," PNAS: 93 (19), 9/17/96, p10268-10273.
Myers (PZ), "Junk DNA is Still Junk," 5/19/10, scienceblogs.com/pharyngula/2010/05/junk_dna_is_still_junk.php
Nakabachi (A.), A. Yamashita, H. Toh, H. Ishikawa, H. Dunbar, N. Moran, & M. Hattori, "The 160-Kilobase Genome of the Bacterial Endosymbiont Carsonella," Science: 314 (5797), 10/13/06, p267.
Namba (Keiichi), "Self-Assembly of Bacterial Flagella," 2002 Annual Meeting of the American Crystallographic Association, San Antonio, TX, www.aip.org/mgr/png/2002/174.htm.
NASA, On the Expansion of the Universe, www.grc.nasa.gov/WWW/K-12 /Numbers/Math/documents/ON_the_EXPANSION_of_the_UNIVERSE.pdf
Nelson (Kevin), Matthew Levy, & Stanley Miller, "Peptide Nucleic Acids Rather Than RNA May Have Been the First Genetic Molecule," PNAS: 97 (8), 4/11/00, p3868-3871.
New Scientist cover story on "The 10 Biggest Mysteries of Life," 9/4/2004.
Nucci (Antonio) & Steve Bannerman, "Controller Chaos," IEEE Sprectrum, 12/07, p43-48.
Nobel-1989,nobelprize.org/nobel_prizes /chemistry/laureates/1989/press.html
Nobel-2004, nobelprize.org/nobel_prizes/chemistry/laureates/2004/illpres/5_proteins.html
Occam's razor, en.wikipedia.org/wiki/Occam%27s_Rasor
Ofria (C.), C. Adami, & T. Collier, "Selective Pressures on Genomes in Molecular Evolution," Journal of Theoretical Biology.: 222, 2003, p477-483.
Ohno (Susumu), "So Much 'Junk' DNA in Our Genome," Brook Haven Symposia in Biology: 23, 1972, p366-370.
OOLprize, Origin of Life Prize, www.us.net/life/
Oparin (A. I.), The Origin of Life, 1952.
Open Letter to Scientific Community, New Scientist, 5/22/04, cosmologystatement.org
Orgel (Leslie), "The Origin of Life on the Earth," Sci Am.: 271, 10/94, p76-83.
Orgel (Leslie), "Self-organizing biochemical cycles," PNAS: 97 (23), 11/7/00, p1253-1257.
Orgel (Leslie), "Prebiotic Chemistry and the Origin of the RNA World," Crit.l Rev. in Biochemistry and Molecular Biology: 39, 2004, p99–123.
Orgel (Leslie), "The Implausibility of Metabolic Cycles on the Prebiotic Earth," PLoS Biol 6(1), 2008, e18.
Orr (H. A.) & Jerry Coyne, "The Genetics of Adaptation: a Reassessment," Am Nat, 1992, p726.
Overbye (Dennis), "Dark energy is Still Puzzle to Scientists," NY Times News Serv, 6/4/08.
Pagels (H. R.), The Dreams of Reason, 1988, p156-58.
Pandolfi (Pier), quoted in "Language of RNA Decoded: Study Reveals New Function for Pseudogenes and Noncoding RNAs," Science Daily, 6/24/10.

Pattee (Howard), "Quantum Mechanics, Heredity and the Origin of Life," J. Theor. Biol. (17), 12/17/67, p410-20.
Paul (Natasha), Greg Springsteen, & Gerald Joyce, "Conversion of a Ribozyme to a Deoxyribozyme Through in Vitro Evolution," Chemistry & Biology: 13, 2006, p329-338.
Pauling (Linus), The Linus Pauling Institute Website, lpi.oregonstate.edu/lpbio/lpbio2.html
Pearson (Helen), "'Junk' DNA Reveals Vital Role," Nature, 5/3/04
Peebles (P. J. E.), & Bharat Ratra, "The Cosmological Constant and Dark Energy," Reviews of Modern Physics: 75, 2003, p559–606..
Penrose (Roger), The Emperor's New Mind: Concerning Computers, Minds, and the Laws of Physics, 1989, p344.
People02, www.prb.org/Articles/2002/HowManyPeopleHaveEverLivedonEarth.aspx
Perkins (D. O.), C. Jeffries, & P. Sullivan, "Expanding the 'Central Dogma': the Regulatory Role of Nonprotein Coding Genes and Implications for the Genetic Liability to Schizophrenia," Molecular Psychiatry: 10, 2005, p69–78.
Piggliucci (Massimo), "Design Yes, Intelligent No," Darwin, Design, And Public Education, 2004, p467.
Pigliucci (Massimo) & Gerd Müller, "Elements of an Extended Evolutionary Synthesis," in Evolution—the Extended Synthesis, 2010, p3-18.
Polanyi (Michael), Quote originally at Michael Polanyi Center site, now at nostalgia.wikipedia.org/wiki/Michael_Polanyi
Poliseno (L.), L. Salmena, J. Zhang, B. Carver, W. Haveman, & P. Pandolfi, "A Coding-Independent Function of Gene and Pseudogene mRNAs Regulates Tumour Biology," Nature, 6/23/10.
Popa (Radu), "A Sequential Scenario for the Origin of Biological Chirality," Journal of Molecular Evolution: 44 (2), 1997, p121-127.
Popper (Karl), "Science as Falsification," Conjectures and Refutations, 1963, p33-39.
Pound (R. V.) & J. L. Snider, "Effect of Gravity on Gamma Radiation," Phys. Rev.: 140(3B), 1965, B788-803.
Prigogine (I.), N. Gregair, A. Babbyabtz, "Thermodynamics of Evolution," Physics Today: 25, 1972, p23-28.
Prigogine (I.), N. Gregair, A. Babbyabtz, Physics Today: 25, 1972, p23-28. Princeton Physics News: 2 (1), 2006, p6.
Provine (Will), "No Free Will," in Catching Up with the Vision, 1999, pS123.
PSSI, Physicians and Surgeons Who Dissent from Darwinism, http://www.pssiinternational.com/list.pdf
Quastler (Henry), The Emergence of Biological Organization, 1964, p16.
Raichle (Marcus) & Debra Gusnard, "Appraising the Brain's Energy Budget," PNAS: 99 (16),

8/6/02, p10237-10239.
Ratzsch (Del), Science & Its Limits, 2000, p123-124.
Ray (C.), "DNA; Junk or Not," The New York Times, 3/4/03.
Ray (T.), "Evolution, Ecology, and Optimization of Digital Organisms," 1992, www.htp.atr.co.jp/~ray/pubs/tierra/tierrahtml.html
Reader (John) & Gerald Joyce, "A Ribozyme Composed of Only Two Different Nucleotides," Nature: 420, 12/19/02, p841-844.
Reviews, Discover Magazine, 12/1/05, discovermagazine.com/2005/dec/reviews
Ridley (Mark), The Cooperative Gene: How Mendel's Demon Explains the Evolution of Complex Beings, 2001, p111.
Ridley (Mat), Genome: Autobiography of a Species in 23 Chapters, 1999, p21-22.
Robertson (Michael) & William Scott, "The Structural Basis of Ribozyme- Catalyzed RNA Assembly," Science:315 (5818), 3/16/07, p1549-1553.
Rode (B. M.), "Peptides and the Origin of Life," Peptides: 20, 1999, p773-776.
Rodin (Sergei) & Andrei Rodin, "Origin of the Genetic Code: First Aminoacyl–tRNA Synthetases Could Replace Isofunctional Ribozymes When Only the Second Base of Codons Was Established," DNA and Cell Biology: 25(6), 6/1/06, p365-375.
Rusbult (Craig), "Methodological Naturalism in Our Search for Truth: A Brief Introduction," www.asa3.org/asa/education/origins/briefmn.htm
Ruse (M.), "Saving Darwinism from the Darwinian s," National Post, 5/13/00, pB.
Ruse (Michael) & E. O. Wilson, "The Evolution of Ethics," in Religion and the Natural Sciences: The Range of Engagement, 1991.
Sagan (Carl), Cosmos, 1960, p18-19.
Sagan (Carl), "Can We Know the Universe?," in Broca's Brain, 1979, p13-18.
Sagan (Carl), Contact: a novel, 1985.
Sagan (Carl) & Ann Druyan, Shadows of Forgotten Ancestors, 1992, p128.
Sagan (Carl), "Life," Encyclopaedia Britannica: 22, 1997, p964-981.
Salisbury (Frank), "Doubts about the Modern Synthetic Theory of Evolution," American Biology Teacher, 9/71, p338.
Salk Institute, "Nuclear Pore Complexes Harbor New Class of Gene Regulators," Science Daily, 2/11/10, sciencedaily.com/releases/2010/02/100204144424.
Schecter (Julie), "How Did Sex Come About?," Bioscience: 34, 1984, p680.
Schiller (F.), "Darwinism and Design Argument," Contempory Review, June 1897.
Schnorrer (F.), C. Schönbauer, C. Langer, G. Dietzl, M. Novatchkova, K. Schernhuber, M. Fellner, A. Azaryan, M. Radolf, A. Stark, K. Keleman, & B. Dickson, "Systematic Genetic Analysis of Muscle Morphogenesis and Function in Drosophila," Nature, 3/11/10.
Schopf (J. William), in Exobiology in the Precambrian paleobiology chapter, 1972, p27-54.

Schultz (Bill), “At the Intersection of ‘Metaphysical Naturalism’ and ‘Intelligent Design’,” 1999, www.infidels.org/library/modern/bill_schultz/crsc.html

Sewell (Granville), “Can ‘ANYTHING’ Happen in an Open System?,” in The Numerical Solution of Ordinary and Partial Differential Equations, 2005, Appendix D.

Serrano (Luis), quoted in “First-Ever Blueprint of 'Minimal Cell” Is More Complex Than Expected," Science, 11/27/09.

SETI Website, setiathome.ssl.berkeley.edu/

Shannon (Claude), “A Mathematical Theory of Communication,” Bell System Technical Journal: 27, July & October, 1948, p379-423 & 623-656.

Shannon (Claude), “Prediction and Entropy of Printed English,” The Bell System Tech. Journal: 30, 1950, p50-64.

Shapiro (James), “Mobile DNA and Evolution in the 21st Century,” Mobile DNA 2010 1:4.

Shrider (Todd), “Evolutionary Control via Sensorimotor Input and Actuation,” ACM Crossroads, Spring, 1998, www.acm.org/crossroads/xrds4-3/evolution.html

Simons (Andrew), “The Continuity of Microevolution and Macroevolution,” Journal of Evolutionary Biology: 15, 2002, p688-701.

Skell (Philip), “Why Do We Invoke Darwin? Evolutionary Theory Contributes Little to Experimental Biology,” The Scientist, 8/29/05.

Smith (John Maynard), Evolutionary Genetics, 1989, p61.

Smith (Wolfgang), “The Universe is Ultimately to be Explained in Terms of a Metacosmic Reality,” in Cosmos, Bios, Theos, 1992, p113.

Smith (John Maynard) & Eors Szathmary, The Major Transitions in Evolution, 1995, p81.

Smolin (Lee), Life of the Cosmos, 1997, p44-45.

Smolin (Lee), The Trouble with Physics, 2007.

Sowerby (Stephen), Corey A. Cohn, Wolfgang M. Heck, & Nils G. Holm, “Differential Adsorption Differential Adsorption of Nucleic Acid Bases: Relevance to the origin of life,” PNAS: 98 (3), 2001, p820-822.

Spetner (Lee), Not By Chance, 1997.

Spies (Maria), “Researchers Probe a DNA Repair Enzyme,” Bio-Medicine, 2/18/08, www.bio-medicine.org/biology-news-1/Researchers-probe-a-DNA-repair-enzyme-2257-1/

Stein (Lincoln), “Human Genome: End of the Beginning,” Nature, 10/21/04, p431.

Steinhaus (H.), Mathematical Snapshots, 3rd ed., 1999, p202.

Stenger (Victor) , God: The Failed Hypothesis: How Science Shows That God Does Not Exist, 2007.

Sternberg (Richard), www.rsternberg.net/smithsonian.php?page=facts, 2008.

Stravropoulos (George), “The Frontiers and Limits of Science,” American Scientist: 65, 11-12/77,

p674.
Superstrings, en.wikipedia.org/wiki/Superstring_theory
Supreme Court Decision, Torcaso v. Watkins (367 U.S. 488), 1961.
Supreme Court Decision, US v. Seeger, 380 U.S. 163, 1965
Supreme Court Decision, Gillette v. U.S., 401 U.S. 437, 450, 1971.
Supreme Court Decision, Harris v. McRae, 448 U.S. 297, 1980.
Sutherland (John), "Ribonucleotides," Cold Spring Harb Perspect Biol, 3/10/10, 2:a005439.
Swee-Eng (A. W.), "The Origin of Life: A Critique of Current Scientific Models," CEN Tech. J.: 10-3, 1996, p300-314.
Swinburne (Richard), "Argument From the Fine-Tuning of the Universe," in Physical Cosmology and Philosophy,1990, p154-73.
Szostak (Jack), "Functional Information: Molecular Messages," Nature, 6/ 12/ 2003, p689.
Tennant (Richard), The American Sign Language Handshape Dictionary, 1998.
Thaxton (Charles), Walter Bradley, & Roger Olsen, The Mystery of Life's Origin, 1992.
Thomas (A.T.), "Developmental Palaeobiology of Trilobite Eyes and Its Evolutionary Significance," Earth-Science Reviews: 71 (1-2), 6/05, p77-93.
Tian (Feng), Owen Toon, Alexander Pavlov, & H. De Sterck, "A Hydrogen- Rich Early Earth Atmosphere," Science: 308 (5724), 5/13/05, p1014 - 1017.
Trefil (James), Harold Morowitz, Eric Smith, "The Origin of Life," American Scientist 97(2), 3-4/09, p206-210.
Trevors (J. T.) & D. L. Abel, "Chance and Necessity Do Not Explain the Origin of Life," Cell Biology International: 28, 2004, p729-739.
Truman (Royal), "Evaluation of Neo-Darwinian Theory Using the Avida Platform," PCID 3.1.1, 11/04.
Turner (Michael), "Hawkings: No miracle in the multiverse," Nature:467, 10/7/10, p657-658.
Tuteja (N.) & R. Tuteja, "Unraveling DNA Helicases, Motif, Structure, Mechanism and Function," Eur J Biochem: 271 (10), 2004, p1849–63.
Valentine (J. W.), et al., "Fossils, Molecules, and Embryos: New Perspectives on the Cambrian Explosion," Development: 126, 1999, p851-59.
Vasas (V.), E.Szathmáry, & M.Santos, "Lack of Evolvability in Self-sustaining Autocatalytic Networks Constraints Metabolism-first Scenarios for the Origin of Life," PNAS, 2/11/10, www.pnas.org/content/107/4/1470.
Veeramachaneni (V.), W. Makalowski1, M. Galdzicki, R. Sood, & I. Makalowska, "Mammalian Overlapping Genes: The Comparative Perspective," Genome Res.: 14, 2004, p280-286.
Venter (Craig), Interview "A Bug to Save the Planet," Newsweek, 6/16/08, p40.
Voet (D.) & J. Voet, Biochemistry, 1995, p1138.
Voie (Albert), "Biological Function and the Genetic Code are Interdependent," Chaos, Solitons

and Fractals: 28(4), 2006, p1000-1004.
Wald (G.), "The Origin of Life" in The Physics and Chemistry of Life, 1955, p12.
Wald (George), "Innovation and Biology," Scientific American:199, 9/58, p100.
Watson (James) & Francis Crick, "Molecular structure of Nucleic Acids," Nature: 171, 1953, p737–738.
Webster's Dictionary for Everyday Use, 1987.
Webster's Third New International Dictionary of the English Language, unabridged, 1993.
Weikart (Richard), From Darwin to Hitler: Evolutionary Ethics, Eugenics, and Racism in Germany, 2004.
Weinberg (Steven), "Life in the Quantum Universe," Articles: Scientific American, proxy.arts.uci.edu/~nideffer/Hawking/early_proto/weinberg.html
Wells (Jonathan), Icons of Evolution, 2000.
Wells (Jonathan) "Survival of the Fakest," The American Spectator, 1/01, p19-27.
Wells (Jonathan), et. al., "The Theory of Intelligent Design: A Briefing Packet for Educators," Discovery Institute, 11/13/07, www.discovery.org/a/4299, p7.
Wickramasinghe (Chandra), Evidence in the Trial at Arkansas, 12/81, www.panspermia.org/chandra.htm
Wiedersheim (Robert), The Structure of Man: An Index to His Past History, 1893.
Wilder-Smith (A.E.), The Scientific Alternative to Neo-Darwinian Evolutionary Theory, 1987, p73.
Williams (Bryony), Claudio Slamovits, Nicola Patron, Naomi Fast, & Patrick Keeling, "A High Frequency of Overlapping Gene Expression in Compacted Eukaryotic Genomes," PNAS: 102 no. 31, 8/2/05, p10936-10941.
Williams (George), "A Package of Information," in The Third Culture: Beyond the Scientific Revolution, 1995, p42-43.
Williams (Peter), (Web) "Intelligent Design Theory – An Overview," www.arn.org/docs/williams/pw_idtheoryoverview.htm
Wilson.(David), "Atheism as a Stealth Religion," Huffington Post, 12/14/07
Wilson (Edgar), An Introduction to Scientific Research, 1990.
Winther (Rasmus), "Systemic Darwinism," PNAS, 8/19/08, p11833-8.
Witzany (Günther), Natural Genetic Engineering and Natural Genome Editing (Proceedings), 12/09, 276 pgs.
Wi-Web, www.wisconsincountyforests.com/qa-forst.htm
Woese (Carl), The Genetic Code, the Molecular Basis for Genetic Expression, 1967.
Woese (Carl), "The Universal Ancestor," Proceedings of the National Academy of Sciences USA: 95, 6/98, p6854-9859.
Woese (Carl), quoted in "Horizontal and Vertical: the Evolution of Evolution," New Scientist,

1/28/10.
Wolfenden (Richard), in "Without Enzyme Catalyst, Slowest Known Biological Reaction Takes 1 Trillion Years," 2003, www.unc.edu/news/archives/may03/enzyme050503.html.
Woodward (Thomas), Doubts about Darwin: A History of Intelligent Design, 2003, p10&20.
Yan (F.), A. Bhardwaj,& M. Gerstein, "Comparing Genomes to Com- puter Operating Systems in Terms of the Topology and Evolution of Their Regulatory Control Networks," PNAS, 5/3/10, 6 pgs.
Yeh (Edward), quoted in "Key Step for Regulating Embryonic Development Discovered," Science Daily, 4/22/10.
Yockey (Hubert), "Self Organization Origin of Life Scenarios and Information Theory," Journal of Theoretical Biology, 1981, p13-31.
Yockey (Hubert), "A Calculation of the Probability of Spontaneous Biogenesis by Information Theory," J. Theor. Biol., 1977, p377–398.
Yockey (Hubert), Information Theory and Molecular Biology, 1992.
Yockey (Hubert), Information Theory, Evolution, and the Origin of Life, 2005.
Yokoyama (Shozo), in "Evolution of a Fish Vision Protein," Physics Today, 9/2/08, http://blogs.physicstoday.org/update /2008/09/evolution_of_a_fish_vision_pro.html
Young (Willard), Fallacies of Creationism, 1985, p165.
Zewail (Ahmed), Nobel Prize in Chemistry, 1999.
Zimmer (Carl), "Testing Darwin," Discover Magazine, 2/5/05, Cover.
Ziv (Jacob) & Abraham Lempel, "Compression of Individual Sequences Via Variable-Rate Coding," IEEE Transactions on Information Theory, 9/78, p530-536.
Zuckerkandl (Emile), "Neutral and Nonneutral Mutations: The Creative Mix-Evolution of Complexity in Gene Interaction Systems," Journal of Molecular Evolution: 44, 1997, p53.

Made in the USA
Charleston, SC
07 May 2011